POLLUTION SCIENCE, TECHNOLOGY AND ABATEMENT

STUDY ON PERSISTENT ORGANIC POLLUTANTS AND THEIR LEGAL GOVERNANCE IN CHINA

POLLUTION SCIENCE, TECHNOLOGY AND ABATEMENT

Additional books in this series can be found on Nova's website under the Series tab.

Additional e-books in this series can be found on Nova's website under the e-book tab.

POLLUTION SCIENCE, TECHNOLOGY AND ABATEMENT

STUDY ON PERSISTENT ORGANIC POLLUTANTS AND THEIR LEGAL GOVERNANCE IN CHINA

YUANYUAN ZENG
QIAN ZENG
AND
KE ZHOU

NOVINKA
New York

Library of Congress Cataloging-in-Publication Data

Names: Zeng, Yuanyuan, (Law teachers) editor. | Qian, Zeng, editor.
Title: Study on persistent organic pollutants and its legal governance in China / Editors: Yuanyuan Zeng, and Zeng Qian, and others.
Description: Hauppauge, New York : Nova Science Publishers, 2016. | Series: Pollution science, technology and abatement | Includes bibliographical references and index.
Identifiers: LCCN 2015043899 (print) | LCCN 2015044611 (ebook) | ISBN 9781634822756 (softcover) | ISBN 9781634842556 ()
Subjects: LCSH: Persistent pollutants--Law and legislation--China. | Environmental monitoring--Law and legislation--China.
Classification: LCC KNQ3131.5 .S78 2016 (print) | LCC KNQ3131.5 (ebook) | DDC 344.5104/633--dc23
LC record available at http://lccn.loc.gov/2015043899

Published by Nova Science Publishers, Inc. † New York

CONTENTS

ABSTRACT

This book includes six chapters. The first chapter introduces the characteristics and classifications of POPs; the second chapter explains the harm of POPs from the perspective of environmental science; the third chapter introduces detailed distribution characteristics of POPs in China; the fourth chapter analyzes the need for legal governance to prevent and control the pollution of POPs in China from the perspective of legislation; the fifth chapter comprehensively reviews the relevant laws, regulations and rules on the prevention and control of POPs in China, at both the macro level and micro level; and the sixth chapter analyzes the problems that arise during the process of legal governance and offers some suggestions from four links, including legislation, enforcement, judicature and law-abiding. The authors hope this research can play a role in the legal governance of POP pollution in China and also that readers will gain an understanding of the state of legal governance of POPs in China. The authors also hope that they give some valued comments for the environmental protection of China.

Keywords: persistent organic pollutants, legal governance, pollution and management

INTRODUCTION

In recent years, with the rapid development of China's economy, the problem of environmental pollution has been increasing significantly. As such, the issue of the legal governance of POPs in China has become an important research subject to which the Chinese government has been paying more and more attention. Based on this, the Chinese government has actively implemented the Stockholm Convention to contain serious situations of POP pollution by enacting a series of policy and legal measures. Although the Chinese government has achieved considerable success, there have been some problems in the process of legal governance.

Chapter 1

INTRODUCTION TO PERSISTENT ORGANIC POLLUTANTS

Persistent Organic Pollutants (POPs) refer to some types of synthetic or natural organic chemicals that can persist in the environment and have harmful effects on human health and the environment by means of the enriched food chains (webs)[1]. Compared with conventional pollutants, POPs are a bigger threat to the natural environment and human health: they exist for a longer period in the natural environment; their molecular structures are more stable and extremely difficult to degrade; they are highly toxic; they can migrate long distances through various environmental media (air, water, organisms etc.) and can spread via global circulation. Once they gain access to an organism via drinking, eating or breathing, POPs are difficult to decompose. They can be concentrated and enlarged along food chains and can cause great harm to humans and animals. Besides having carcinogenic, teratogenic and mutagenic characteristics, some POPs also have endocrine-disrupting effects. The effects of POPs on humans and the environment can be sustainable for several generations. They represent a major threat to human survival and sustainable development.

---

[1] Yu Gang. Persistent organic pollutants: a new global environmental issue [M]. Science Press, 2005.

1.1. THE CHARACTERISTICS OF POPS

As with other organic pollutants, when POPs enter the environment they cause a series of physical, chemical and biological reactions, such as distribution, adsorption, volatilization, bioaccumulation, oxidation, hydrolysis, photolysis, and biotransformation. However, due to their own unique characteristics, POPs are different from general organic pollutants. These unique features of POPs include persistence, bioaccumulation, semi-volatility and high toxicity.

1. Persistence

Most POPs have stable molecular structures; they are extremely difficult to degrade under natural conditions, and they have a strong ability to resist common metabolic means, such as biological metabolism, light degradation, chemical decomposition and so on. Once released into the environment, POPs are difficult to decompose; they can be stable and continuously accumulated in the environment, and they can be residual in water, soil, air and sediment for several years or longer.

We often use the half-life period ($t_{1/2}$) as a measure of POPs' persistence evaluation parameters. For example, the volatilization's half-life period is defined as the required time for the concentration of volatile pollutants to reduce by half. Usually, POPs' $t_{1/2}$ values in water, sediment and soil are more than 60d, 180d and 180d[2].

2. Bioaccumulation

Bioaccumulation is a phenomenon whereby organisms in the process of growth and development absorb and accumulate external matter directly from the environment or from the consumption of food.

The molecular structure of POPs usually contains a halogen atom, which is more soluble in organic solvents than in water. It can bioaccumulate in adipose tissue when it enters the body of a human or other organisms via

[2] Hou Deshen. Study on the distribution characteristics and ecological effects of persistent toxic substances in the western region of Inner Mongolia [D]. Mongolian University Press, 2014.

water, food or air. It causes POPs to enter an organism from the surrounding matter; it abounds in animals that stand in the high-grade of the food chain, and ultimately reaches toxic concentrations.

The mechanism of bioaccumulation is a distribution process of organic compounds in the fat/water system. Bioconcentration is usually denoted by the bioconcentration factor (BCF). BCF is the ratio of organic compounds' concentrations in organics or biological tissues, as well as in water. It is a measure used to estimate the capacity of bioaccumulation of aquatic organisms. As aquatic organisms in nature, the number of fish species and their populations are very large, so they are important food sources for humans and other carnivorous animals. Therefore, fish can be used as an effective standard test of BCF.

The study found that BCF and the octanol/water partition coefficient of organic pollutants are strongly correlated[3]. Kow is one of the important physical and chemical property parameters of compounds. It is the ratio of the compound's amount of substance concentration into balance in octanol phase and water phase.

$$Kow = [Co]/[Cw]$$

Co- concentration of organic matter in the octanol phase, mol/L;
Cw- concentration of organic matter in the water phase, mol/L;

Usually, Kow is measured at room temperature (25^0C) and the total concentration of organic matter does not exceed 0.01mol/L.

People often use the logarithmic form lgKow for convenience in real life. Organic lgKow ranges from hydrophilic compounds -4.0 to 8.5 of hydrophobic compounds. Due to POPs' stronger lipotropy nature, its lgKow is greater than 4.0, in general. In addition, lgKow also has an obvious correlation with numerous biological activities of organics, such as the variety of pharmaceutical functions, toxicity, insecticidal activity and so on. It also has important significance as a lipophilic or hydrophobic measure parameter to elaborate the interaction between organics and macromolecular matter or receptors.

[3] Zhao Yanyan, Wang Ling, Lou Yinghua, et al. Determination of a novel bromation flame retardant in the intertidal sediments on the north coast of Jiaozhou Bay by GC-MS[J]. Environmental Chemistry, 2015(2).

Meanwhile, in the study of the bioaccumulation process of POPs, the solubility of organic matter is closely related to Kow in water; it is also an important parameter that must be considered.

In addition to these parameters, the organic carbon distribution coefficient (Koc), soil/sediment distribution coefficient (Kd), and other parameters can also be used as important evaluation indices in researching the environmental behavior and toxicity of POPs.

3. Semi-Volatility

All POPs have semi-volatilities, meaning that they can volatilize into the atmosphere from the water or soil, and they are usually adsorbed by atmospheric particles to migrate long distances in the environment via atmospheric circulation. At the same time, due to their semi-volatility, POPs cannot exist in the air perpetually. When the temperature decreases or the particles happen to sedimentate, POPs can settle to the surface of the earth, and then move into soil and water. At the same time, this process can repeat itself several times. Due to their persistence and semi-volatility, POPs can be monitored in all ecological environments, including mainland desert seas and the South Polar Region.[4] Research shows that even in the sparsely populated Arctic, high concentrations of POPs have also been detected in the bodies of living mammals. The data from such detection includes: 2~39000 (μg/g wet weight) DDT, 300~118000 (μg/g wet weight) Dioxins, 0.6~7096 (μg/g wet weight) Chlordane, 1~12900 (μg/g wet weight) PCBs, and 84~9160 (μg/g wet weight) Toxaphene.

The parameters of POP volatility mainly include saturated vapor pressure (Ps, Pa) and the Henry constant (KH). Vapor pressure is the degree at which chemicals move from solution or solid into space. Since Ps is constant at a certain temperature, it is generally used as an evaluation of POPs volatility evaluation parameters. Generally speaking, compounds with low relative molecular mass (Mr) have higher saturated vapor pressure; POPs' Mr is middling, so they exhibit semi-volatilities. In addition, KH can also be used to indicate the size of the volatile compounds. When the compounds in air and water achieve a relative balance, it is the ratio of the vapor pressure and water

[4] Hu Riledawa. Distribution characteristics of OCPs, PCBs and heavy metals in soil, grass and parks in Mongolia [D]. Mongolian University Press, 2014.

solubility of compounds under standard temperature and pressure (0^0C, 100kPa), and its unit is Pa•m^3/mol or atm•m^3/mol.

4. High Toxicity

Most POPs have high toxicity, and some also have carcinogenicity, teratogenicity, mutagenicity, reproductive toxicity and immune toxicity. Both laboratory studies and epidemiological investigations have shown that POPs can cause endocrine disorders, reproductive and immune system dysfunction, and even renal failure, cancer, or other serious diseases in organisms. Also, the toxicity may continue for some time due to the pollutants' persistence.[5]

After POPs enter the organism, their toxic effects can be broadly divided into two types:

1) The toxicity of POPs that comes from their chemical structure is equal to the substance's physiological action. Due to the substance's concentration, its physiological action is proportionate to the substance's amount in the organism. In other words, when POPs' concentration is extremely low, they show no effect; this is the so-called no-effect level. However, when the concentration exceeds a certain threshold, the inherent physiological action begin to occur[6]; if the concentration continues increasing to reach a lethal dose, the organism will not maintain normal physiological metabolic function, which eventually leads to death. If POPs enter the organism followed by biological metabolism and transformation, and are then expelled from the body by the substance's polar interaction, resulting in the decrease of POPs' concentration in organism, then the symptoms of poisoning have been shown to improve; that toxicity is known as Uncomplicated Acute Toxicity.

2) After entering the organisms, POPs can generate unstable intermediates that have strong abilities in the biological metabolizing enzymes and polarization process. Also, a portion of POPs could covalently bind to the cell polymer composition, including proteins and nucleic acids, resulting in irreversible chemical modifications. Protein modification can lead to tissue necrosis and allergic reactions, but the chemical modification of nucleic acid

---

[5] Ajay Kumar Goel. Anthrax: A disease of biowarfare and public health importance [J]. World Journal of Clinical Cases, 2015(1).

[6] Rui Chen, Qing Yan Xu, Bai Cheng Liu. Simulation of the Dendrite Morphology and Microsegregation in Solidification of Al–Cu–Mg Aluminum Alloys [J]. Acta Metallurgica Sinica, 2015 (2).

can damage the normal information transmission of cells; it can cause cell death or mutation, and can result in tumors of the tissue.

In addition, there are some POPs in organisms that can change into other substances that are more poisonous, which results in toxic effects on organisms. Let us take heptachlor as an example; it can be turned into heptachlor epoxide in soil and organizations of plants and animals, where it is 4 times more toxic than the original compound.

In order to describe the POPs' behavior in the environment, it is common to use $t_{1/2}$, Ps, KH, BCF, Koc, Kow and other parameters to show the migration and transformations' regular pattern of these materials. These parameters are important indicators for understanding the POPs' rules of migration and transformation. They are also important indicators for monitoring the degree of pollution.[7] Table 1-1 lists the parameters of the characteristics of POPs. The temperature, which is not indicated in the table, is 25°C.

1.2. The Classification of POPs

There is a huge number of homologous compounds and isomers of POPs in chemical structures - over tens of thousands, in fact.[8]

1.2.1. POPs' List of International Conventions

The Stockholm Convention on persistent organic pollutants aims to regulate the first controlled list of POPs, consisting of a dozen varieties.[9] Generally speaking, 12 kinds of POPs can be divided into three categories, or three sources.

The first kind is organic chlorine pesticides (fungicides): Aldrin, Chlordane, DDT, Dieldrin, Endrin, Heptachlor, Mirex, and Toxaphene.

---

[7] Yajun An, Fang Huang. A Review of Mg Isotope Analytical Methods by MC-ICP-MS[J]. Journal of Earth Science, 2014-10-15.

[8] Gai Nan. Ecological geochemistry study of organochlorine pesticides and polychlorinated biphenyls on Ruoergai wetland[D]. China Academy of Geological Sciences Press, 2012-06-08.

[9] Allan Astrup Jense. Why has the Stockholm Convention Banned the Brominated Flame Retardant Hexabromocyclododecane (HBCDD)[C]. 2014-9-21.

Table 1-1. 12 kinds of characteristic parameters of POPs

Name	half-life $t_{1/2}$	Ps(liquidoid)/ Pa	KH/(Pa·m^3/mol)	lgBCF	lgKoc	lgKow
Aldrin	20~100d (soil) 35min (air)	$1.87*10^{-2}$	91.23	5.40	2.61-4.69	5.17-7.40
Dieldrin	5a (soil), 4months (water)	$0.72*10^{-3}$	5.27	3.30-4.50	4.08-4.55	3.69-6.20
Endrin	14a (soil), 4a (water), 5-9d (air)	$0.27*10^{-4}$	0.64	3.82	3.23	3.21-5.60
DDT	100d (volatilization from the soil surface), >150a (water), 2d (air)	$0.21*10^{-4}$ (20°C)	0.84 1.31(23°C)	3.70	5.15-6.26	4.89-6.91
Chlordane	-la (soil)	$1.33*10^{-3}$	4.86	5.56	4.58-5.57	6.00
Heptachlor	2a (soil), 1d (water) 36min (air)	$4.0*10^{-2}$	$2.33*10^{2}$	4.59	4.38	4.00-5.50
Mirex	10.7h (water)	$1.07*10^{-4}$	82.17	6.20	7.38	5.28-6.89
Toxaphene	1-14a (soil), 6h (water), 4-5d (air)	26.7-53.3	$6.38*10^{3}$	3.59	3.18	3.23-5.50
HCB	-4a (soil), 8h (water), 2a (air)	$1.45*10^{-3}$ (20°C)	$7.2x10^{2}$ (20°C)	6.40	2.56-4.54	3.03-6.42
PCBs	>6a (soil), 21d-2a (air) (-$1.01*10^{4}$ (except for a chlorine and dichloro-diphenyl))	$2.1\text{-}4.0*10^{-7}$	$1.01*10^{3}$	4.00-5.00	5.49	4.30-8.26
PCDDs	2weeks-6a (soil), 8weeks-6a (sediment) - $1.7*10^{-2}$, 2d-8weeks (water), 2-21d (air)	$1.1*10^{-10}$	0.13-3.34	45.47	4.36-7.81	4.30-8.20
PCDFs	8weeks-6a (soil), 2weeks -6a (sediment), 3-8weeks (water), 1-3weeks (air)	$5.0*10^{-10}$~$3.9*10^{-4}$	0.42-1.50	3.00-4.00	3.92-7.61	5.40-8.00

The second kind is industrial products; PCBs and HCB.

The third kind comprises of chemical impurities and their derivatives: PCDDs and PCDFs.[10] This kind of matter did not exist in nature, which has caused great concern in recent years. They are simply a class of substances produced by industry or other human activities. Their creation is mainly due to the combustion of fuel, including some emissions from garbage incinerators and medical waste combustion furnaces.

12 kinds of molecular formulas of organic pollutants and their sources are listed in Table 1-2.

Table 1-2. 12 kinds of molecular formulas of organic pollutants and their sources

Name	Molecular formula	Effects and Sources
Aldrin	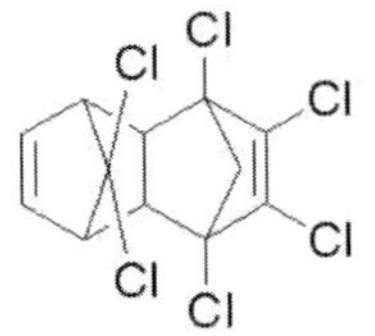	pesticide
Chlordane	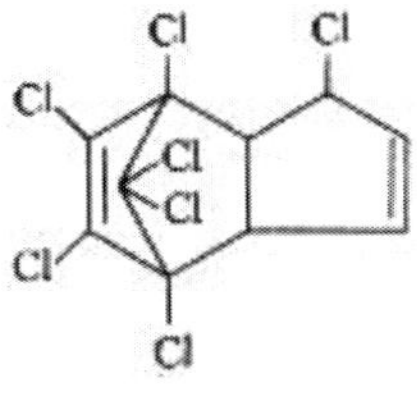	pesticide
DDT	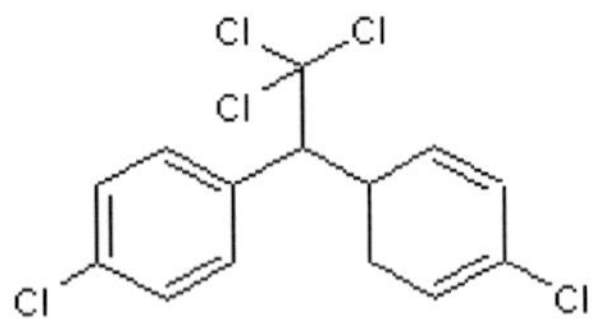	pesticide

[10] Wen Sheng, Hui Yang, Yang Fangxing. Polychlorinated dibenzo-p-dioxins (PCDDs) and dibenzofurans (PCDFs) in surface sediment and bivalve from the Changjiang Estuary, China [J]. Chinese Journal of Oceanology and Limnology, 2008-03-15.

Name	Molecular formula	Effects and Sources
Dieldrin	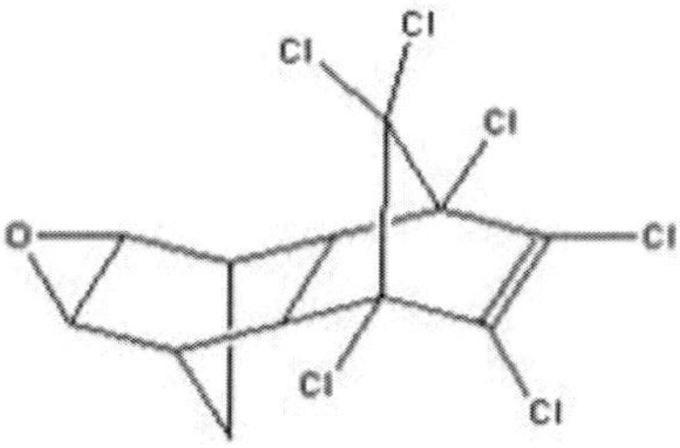	pesticide
Endrin	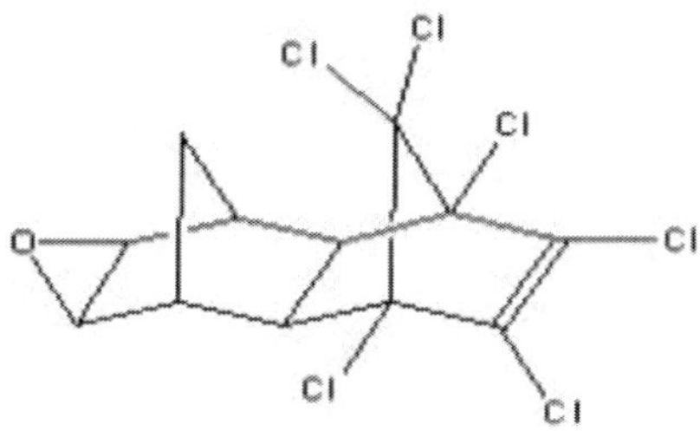	pesticide
Heptachlor	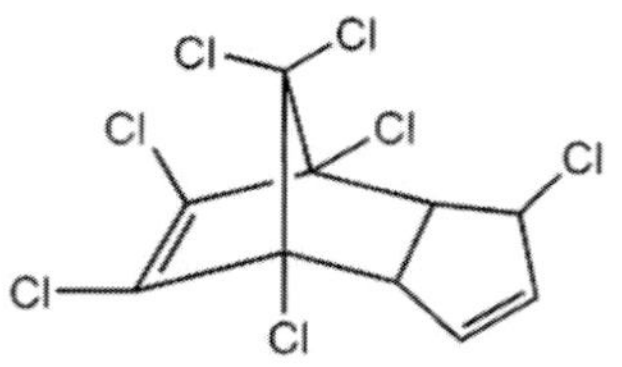	pesticide
Mirex	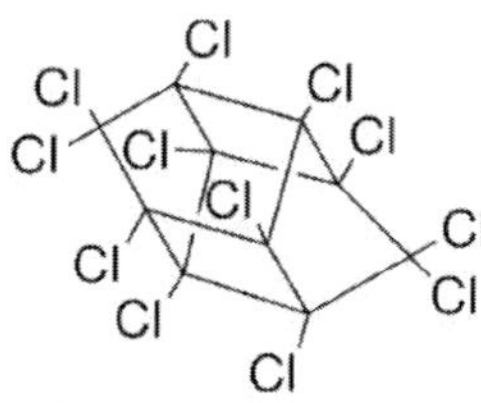	pesticide
Toxaphene	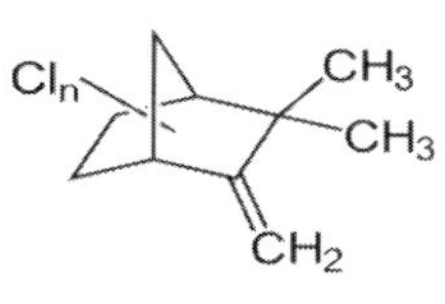	pesticide

Table 1-2. (Continued)

Name	Molecular formula	Effects and Sources
HCB	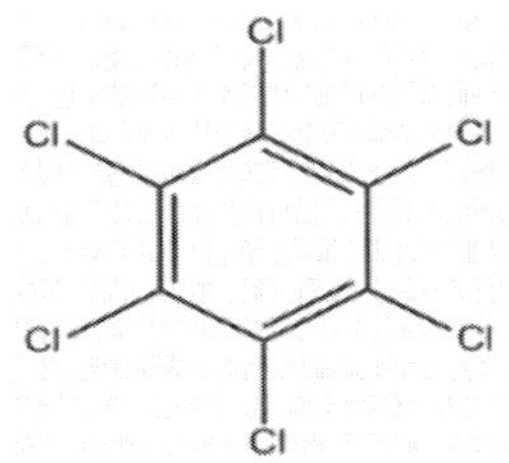	fungicide
PCBs	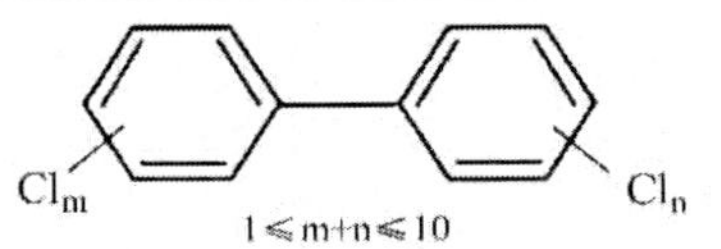	Plasticizers, lubricants, electrolytes, and other fine chemical products
PCDDs	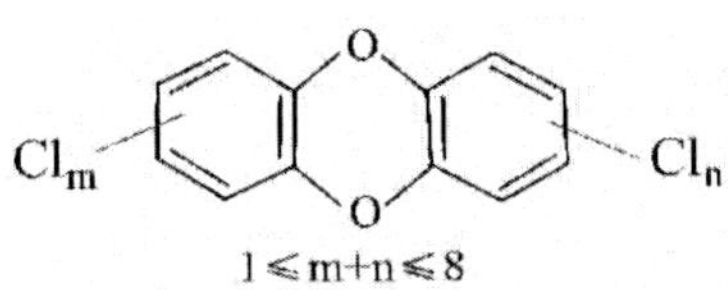	Chemical derivatives, industrial intermediate products, waste combustion byproducts, chemical products' impurities derivatives
PCDFs	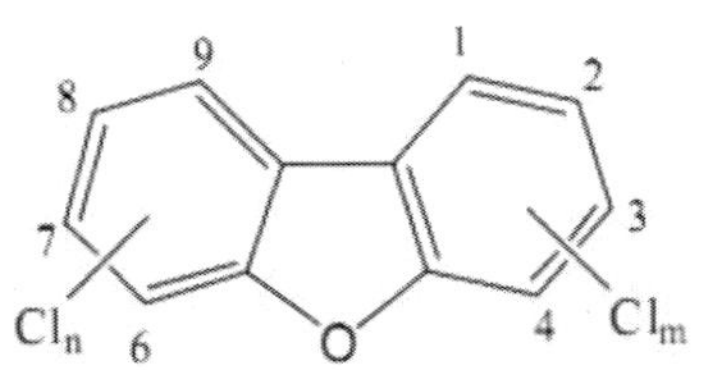	Chemical derivatives, industrial intermediate products, waste combustion byproducts, chemical products' impurities derivatives

1. Aldrin

Aldrin has been used to control warehouse pests, termites and so on, and has a carcinogenic effect on humans.[11]

2. Dieldrin

Dieldrin has been used to control pests and murine in corn, rice, cotton and sugar cane, and has carcinogenic effects on humans.

[11] Liu Wenmin. Effect of PBDEs on expression of genes related to the immunity of zebrafish [D]. Shandong Normal University Press, 2014-03-26.

3. Endrin

Endrin has been used to control pests and murine in corn, rice, cotton and sugar cane, and has carcinogenic effects on humans.

4. DDT

Since its invention in 1939, the broad-spectrum, high-residual pesticide DDT has played a significant role in the elimination of agricultural and forestry pests, and won the Nobel science award. Due to being fat-soluble, it is easy to accumulate in animal fat and extremely slow to catabolize. When it is stored in the body at certain levels, it will damage the central nervous system, liver, and thyroid, ultimately leading to death.

5. Heptachlor

Heptachlor has effects on contact, stomach poison and fumigation, and is mainly used for killing pests. It is harmful to the body's immune and reproductive systems, and also has a carcinogenic effect.[12]

6. Chlordane

Chlordane has effects on contact, stomach poison and fumigation, and is mainly used in the prevention of termites in buildings. It also harms the immune system. Although our country has banned its production, chlordane is still used and stored by some people.

7. Mirex

Mirex is a chronic stomach insecticide that has significant effects in the prevention of ants and termites. After spraying it on an insect population natural communication, carcinogenic effects are found in humans after 15 days. It is understood that, at present, there are still some units and individuals who mass produce, use, and store Mirex. They also process, produce and sell termite control products and health pharmacy. People should be cautious when buying or using such products.

[12] Wang Chunhui, Wu Shaohua, Zhou Shenglu, et al. Research progress on the characteristics and environmental behavior of typical soil POPs [J]. Environmental Chemistry, 2014-11-15.

8. Toxaphene

The broad-spectrum, high-residual pesticide Toxaphene is four times more powerful than DDT, which can cause thyroid tumor and cancer.

9. HCB

HCB fungicide is commonly used in controlling fungal diseases in crops, but it harms the immune and reproductive systems of humans.[13]

10. PCBs

Polychlorodiphenyls (PCBs) have been used in the plastic industry and in the production of transformers, capacitors, paint and paper. They damage the immune and reproductive systems of animals.

11. Dioxins (PCDDs, PCDFs)

Dioxin is the most toxic organic pollutant of all environmental pollutants. It comes from the bleaching of paper, the combustion of gasoline, old metal recycling, organic chlorine synthesis and other organic chemical manufacturing processes. Volcanic activity and forest fires can also form dioxins. Dioxin mainly enters humans and animals through the food chain, as well as through skin contact and breathing. Dioxins in the body are mostly accumulated in the liver and adipose tissue. In 1997, international cancer research institutions issued a statement confirming dioxin as a human carcinogen. Human epidemiological research also confirmed that the load of dioxin in the body was positively related to cancer incidence: when the body fat's dioxin load goes up to five ppm, there will be 0-40 cancer patients in every one million people.[14] In addition, it can also damage the reproductive and endocrine systems.

[13] He Shichang. Chemical characteristics of organochlorine pesticides in Antarctic route atmosphere [D]. Third Institute of Oceanography of State Oceanic Administration, 2013.

[14] Yang Qiaoyun. Study on the exposure and health effects of typical organic pollutants in the body [D]. Peking University, 2014-06-01.

Table 1-3. LRTAP new molecular formulas and sources of POPs

Name	Molecular formula	Effect and source
HCH	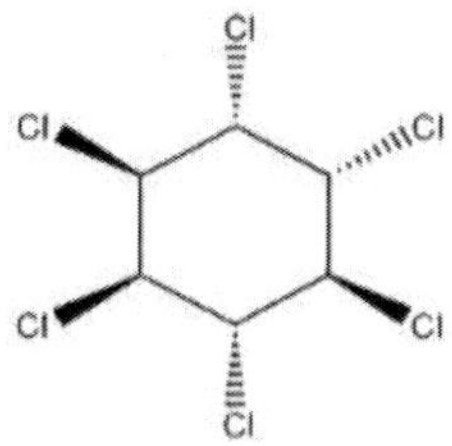	pesticide
Lindane	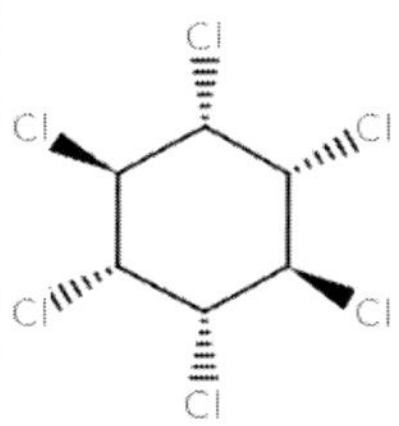	pesticide
PAHs	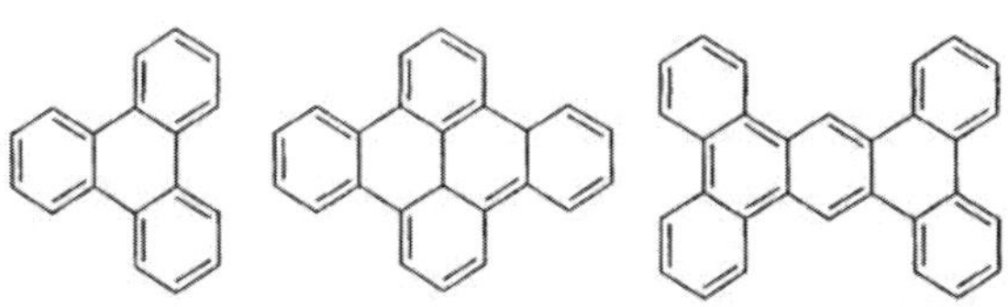 三亚苯 triphenylene　二苯并[e,i]芘 dibenzo(e, i)pyrene　四苯并[a, c, h, j]蒽 tetrabenzo(a, c, h, j)anthracen	Biosynthesis, video production, coal and oil, transport emissions, waste incineration and other harmful impurities derivatives
PCP	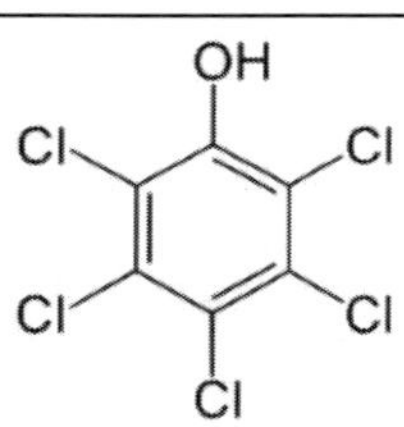	Traditional preservative for textile, leather, wood, slurry

Table 1-4. Blacklist of priority pollutants of China

Chemical category	Name
1. Chlorinated hydrocarbon	methylene chloride, chloroform, carbon tetrachloride, chloroform, methylene chloride, Carbon tetrachloride, trichloroethylene, tetrachloroethylene, bromoform, 1,2-dichloroethane, 1,1,1-trichloroethane, 1,1,2-trichloroethane, 1,1,2,2-tetrachloroethane
Chemical category	Name
2. Benzene series	Benzene, toluene, ethylbenzene, ortho-xylene, meta-xylene, para-xylene
3. Chlorophenyl-glycine	Chloroebenzene, Ortho-dichlorobenzene, para-dichlorobenzene, hexachlorobenzene
4 PCBs	polychlorobiphenyl
5. Phenols	Phenol, meta-toluene, 2,4-dichlorobenzene, 2,4,6-trichlorobenzene, pentachlorobenzene, para-nitrophenol
6. Nitrobenzene	Nitrobenzene, para-nitrobenzene, 2,4-dinitrotoluene, trinitrotoluene, para-nitrochlorobenzene, 2,4-dinitrochlorobenzene
7. Phenyl amines	Aniline, dinitraniline, para-nitroaniline, 2,6-dichloronitrobenzene amine
8. Polycyclic aromatic hydrocarbon (PAHs)	Naphthalin, fluoranthene, benzo[b]fluoranthrene, benzo[k]fluoranthrene, benzo[a]pyrene, Indeno[1,2,3-cd]pyrene, benzo[ghi]pyrene
9. Phthalic ester	dimethyl phthalate, dibutyl phthalate, dioctyl phthalate
10. Pesticides	Hexachlorocyclohexane, dichlorodiphenyl trichloroethane (DDT), panaplate, Rogor, parathion, methyl parathion, nitrofen, dipterex
11. Acrylonitrile	acrylonitrile
12. Nitrosamines	N-nitroso dipropyl amine, N-nitroso di-n-propylamine
13. Cyanide	cyanide
14. Heavy metal and its compound	Arsenic (As) and its compound, Berillium (Be) and its compound, Cadmium (Cd) and its compound, Chromium (Cr) and its compound, Copper (Cu) and its compound, Lead (pb) and its compound, Mercury (Hg) and its compound, Niccolum (Ni) and its compound, Thallium (Tl) and its compound

1.2.2. POPs List of Regional Conventions

America, Canada and 32 European countries signed the "Convention on Long-Range Transboundary Air Pollution (LRTAP)" in the Pan-European environmental ministers' meeting held in June 1998 in Aarhus, Denmark.[15] The convention presents 16 (class) POPs that need to be controlled. Besides the 12 kinds listed by the United Nations Environment Program (UNEP), the list includes another four kinds of POP: HCH, Lindane (99.5% HCHγ Hexachlorocyclohexane agent), PAHs and PCP (Table 1-3).

1.2.3. New List of POPs

Some agencies and non-governmental organizations have successively put forward proposals about the new POPs. In August 2004, the EU suggested enlarging the list of POPs in a press release that said "Chemical pollution: The Committee would like to remove more dirty materials from the world." It added nine new kinds of POP: kepone, hexabromobiphenyl, hexachloro-cyclohexane, PAHs, hexachlorobutadiene, Eight polybrominated diphenyl ether, pentachlorobenzene, polychlorinated naphthalene (PCN), and Short Chain Chlorinated Paraffins (SCCP).

In addition, some academic and non-governmental organizations came up with new classes of POPs, including chlorpyrifos, atrazine and Perfluorooctane sulfonate (PFOS). We also need to pay attention to certain carcinogens, such as Azaheterocyclic (N, S, O, P, X) compounds, metallo-organic compounds, including methylmercury, aluminum methyl and tin trimethyl $Sn(CH_3)_3$, animal drugs, propylene chlorohydrin and acrylamide. Most of the major organic pollutants of China are POPs (Table 1-4).

[15] Hong Bing. The study of POPs in Chinese alligator [D]. Anhui Normal University Press, 2014.

Chapter 2

THE HARM OF PERSISTENT ORGANIC POLLUTANTS

POPs are chemical substances synthetized by humans; they can exist in the environment sustainably, accumulate through the biological food chain (net) and are detrimental to human health. They have four characteristics: they have high toxicity, they are persistent, biologically cumulative, and lipotropy hydrophobic, but at the top of the biological chain, humans have increased the toxicity to 70 thousand times. POPs' special properties make them extremely harmful to human health and the ecological environment. The study found that POPs not only have high toxicity, but are also proven to be an origin of disease and damage to humans, animals, and plants. Once the POPs enter the organisms, in any number of ways, they will be accumulated in fat, embryo liver and other organs. When accumulated in a certain concentration, they will cause harm to organisms. Not only do they cause chronic and acute toxicity to the liver, kidney and other organs, as well as the nervous, endocrine, and reproductive systems, but they also possess carcinogenicity, genotoxicity, neurotoxicity, and endocrine disruptors. POPs in the environment have posed a serious threat to the ecological environment and human health.[16]

From 1950 to 2000, the global environment was polluted by a large number of POPs. The environmental pollution events related to POPs began to emerge in an endless stream. For example, in 1968, the Kyushu residents of Japan ate rice bran oil containing PCB, and thousands of people were

[16] ten Tusscher G W, KOPPE J G. Chemosphere, 2004, 54: 1329-1336.

poisoned. Taiwan also experienced a similar event in 1979.[17] America used a defoliant containing PCB during the Vietnam War, leading to health problems for the local people, U.S. soldiers, and their offspring. In July 1976, Iraq Kemosa chemical company bloomed and nearly 2KG dioxin POPs were released. This incident caused the deaths of many rabbits, birds and mice in Seveso town. Many children's faces developed water marks and more than 700 people were compelled to leave their homes. From 1998 to 1999, many people in Germany, Belgium and other countries experienced fowl poisoning due to PCB and dioxin. It was spread globally, causing huge economic losses.[18,19] POP pollution in the global environment presented a big challenge for decision makers and environmentalists. The challenge consisted of deciding how to describe these global pollutants' features and understand POPs' toxic mechanism. Based on this, this chapter mainly covers the research on POPs' harmful aspects to ecosystems and human health.

2.1. POPs' Toxic Mechanism

POPs' toxic mechanism for biology is not completely obvious. When POPs cause harm to human health, it is generally not one or single kinds of POPs, but the result of interactions between several POPs. All of the pollutants that enter the environment are produced by human activities, but POPs represent the biggest threat to human life. They can cause endocrine disturbances, destroy reproductive and immune systems, lead to cancer, cause malformations, genetic mutations and nervous system diseases. Also, they can remain in the body for several weeks, which is a serious threat to human survival and the sustainable development of society.

Xenobiotics' toxic mechanism to organisms is diverse. Complex toxic mechanisms may be involved at multiple levels and steps. First, the poison is sent to the target site, where interaction and combination occurs, due to the disorder of cell function and structure, which leads to repair at the cellular or molecular level. When the target molecular mechanism of the toxin induces changes in the functional disorder or the ability to repair itself faces obstacles, namely toxicity, tissue necrosis or fibrosis, cancer and other toxic effects result

[17] Yu Gang. Persistent organic pollutants: a new global environmental issues [M]. Beijing: Science Press, 2005.

[18] Frignani M, Bellucci L G, Carraro C et al. Chemosphere, 2004, 54: 1563-1572.

[19] Jones K C, de Voogt P. Environ. Pollut., 1999,100: 209-221.

and develop in a variety of ways. In general, there are several main theories to explain the role of the toxic substance on the target site (molecule).[20]

1. Target Theory

When pollutants (namely the ultimate toxicant) reach a certain concentration in the organism, a series of harmful biological effects are caused at the site, called the target site. The chemical must reach the target organ and reach a high enough concentration to directly play a role; the concentration may be higher or lower than the blood concentration.

2. Covalent Binding Theory

This theory suggests that the toxicant and target molecule are combined by covalent and non-covalent forms. Non-covalent binding is reversible and relatively low energy; covalent binding is non-reversible and can permanently alter endogenous molecules.

3. The Theory of Free Radicals

Free radicals are the unpaired electrons of atoms or molecules that mainly form due to the energy of covalent bond homolysis, and can also be produced by electron capture. There are three ways that pollutants become free radicals: the formation of free radicals in the environment through pyrolysis and combustion, photolysis, and oxidation-reduction reaction processes; through a biological transformation process that forms various free radical intermediates, and in biological macromolecules, covalent bond homolysis occurs in the body, which generates free radicals; there are also many organic toxicants (such as PCBs) that can produce free radicals during body metabolism.

[20] Huijuan Xiu. Environmental toxicology[M]. Beijing: Environment Science Press of China, 2003.

4. Receptor Theory

According to Receptor theory, the receptor is the large protein present in the cell membrane which has the ability to identify pathognomonic bioactivators and selectively bind with them. A variety of receptors have high sterospecificity and are just a small part of the cells, and the quantity is limited; receptor proteins also possess synthesis and degradation processes, and the number is not fixed. The receptor can be affected by many internal and external factors, including pollutants.

The toxicity mechanism also relates to the cell repair function's imbalance. Many toxins can cause changes to the structure of biological macromolecules. If the damage is not repaired in time, it will cause damage to organisms.

2.2. Detriment to the EcoSystem from Persistent Organic Pollutants

By entering the environment in various ways, and subsequently spreading and accumulating throughout food chains, POPs can cause severe damage to the ecosystem. POPs accumulating in adipose tissue in organisms will influence embryogenesis via the placenta and lactation, leading to malformation, stillbirth, EUGR etc. When exposed to high concentrations of POPs, the spawning rate of birds will modestly decrease, and the population quantity will constantly reduce, which could reach extinction levels. In an experiment on POPs toxicity towards chickens, PCB was found to induce death in chicken embryos and edema to various degrees, and further boost the death rate of hatching eggs. Due to these reasons, osprey and eagles are almost extinct in some regions of the USA.

A high level of POPs' existence and use will also do tremendous harm to the agriculture system.[21] [22] For example, in the mid-1980s Indonesia, one of the largest rice producers in the world, heavily used OCPs, which resulted in the extinction of all natural enemies of locusts, finally leading to an unprecedented disaster of locusts, whereby a huge amount of rice was destroyed.

[21] Liu Ling, Chui Guangbo, Hao Zhenchun. Journal of hydraulic engineering, 2003, 6:6-13.

[22] Boening D W. Environ. Safety, 1998, 39:155-163.

After accessing organisms via many approaches, POPs will accumulate in adipose tissue, embryos, the liver and other organs, and cause significant harm to the body when they accumulate to a certain level. After the 1990s, numerous research studies into POPs' toxicity and ecosystem influence were conducted in Europe and the Great Lakes Region. The dysgenesis of fish birds in Europe and the Great Lakes Region was caused by POPs, which gave rise to beluga in Lawrence dysgenesis. Furthermore, research suggested that the influence of POPs had expanded to the immune systems of top predators, proving the pathopoiesia possibility for humans and their effect on animal behaviors. The most serious damage was to the top predators in the food chain, including: 1) dysgenesis and population decrease; 2) dysfunction and hormonal abnormality; 3) gender-variance; 4) immune system disorders; 5) behavior disorders; and 6) tumors and cancer. http://dict. youdao.com/ javascript:void(0).

POPs and heavy metals are the primary factors that contribute to many diseases and damage to living things. They impact people mainly through food chains, and secondly through the air and skin contact. Often, POPs are first absorbed by plants, marine microorganism and insects, and then move into food chain cycles, eventually contaminating fish, meat and dairy products. When people eat food contaminated by POPs, they will be stored in their fat fiber, and can then be passed to infants through the placenta and lactation. The damage to humans usually comes in four types. The first kind is the impact on the newborn's weight, which may decrease, dysplasion, and disorders of skeletal development and metabolic activity. The second kind is the oppression of the nervous system, attention disorders, and impacts on the immune system. The third kind is the harm caused to the reproductive system. Besides, it also potentially threatens the human endocrine system. For men, POPs can lead to testicular cancer, the reduction of sperm, reproductive dysfunction, and an imbalanced sex ratio in newborns. For women, POPs can lead to breast cancer, the early onset of puberty and so on. POPs not only cause harm to individuals, but also have permanent effects on their offspring. The fourth kind is the development of cancer.

2.3. The Damage that Persistent Organic Pollutants Cause to Human Health

POPs are a global environmental issue. Their danger to human health is becoming more and more obvious. Recently, the morbidity of cryptorchidism, hypospadias, endometriosis, bisexual issues, and aplasia is increasing, and teenage girls' adolescence is ahead of schedule. These are all considered to be connected with the pollution of environmental hormones (including POPs).

Although there has been recent worldwide controversy about the reports of male semen and the density of semen and sperm counts, in the past few decades there has been a trend in some areas for male semen and sperm counts to decrease. These situations may even lead to malignant tumors in the reproductive system. Thus, some people think that POPs are the true time bomb threatening human survival.

Therefore, various industrial and developed countries, including America, Japan, and many countries in Europe, have begun to pay more attention to research on POPs.

In China, with its rapid economic development, environmental problems are growing increasingly serious. In Beijing, a study of persistent organic pollutants (POPs) found that 90% of 300 pregnant women in Beijing had polychlorinated biphenyls (PCBS) or machine phosphorus pesticides in their breast milk. 10% of those were at a dangerous level. As such, environmental problems in China are gaining more attention. What's more, the research into POPs attracts attention from all walks of life, as it is truly a global issue.

Chapter 3

THE DISTRIBUTION CHARACTERISTICS OF POPs IN CHINA

POPs have the characteristic of long retention in a natural environment, structural stability, low degradability, intensive toxicity and the ability to spread via a global cycle. Due to their characteristics, they have been investigated in the atmosphere, waters, sediments and soils of China. These substances are assimilated by organisms via drinking, feeding, and respiration and thus result in bio-magnification owing to their tough degradability, which poses potential threats to human health in China.

There are 6 main kinds of OCPs in practical use and production in China, including DDT, HCB, Heptachlor, Toxaphene, HCH, and Chlordane,[23] while 4 other OCPs (Aldrin, Dieldrin, Endrin, and Mirex) are not officially being produced. At present, there is still a small amount of production of chlordane, HCB and DDT. Among all types of OCPs, HCB accounts for the largest amount, and is far more prevalent than other OCP outputs. According to Yang Xiaomei's study,[24] DDT and HCB were detected to be the major part of OCPs in different media in China, to varying extents, including air, waters, sediments and soils. Soil pollution is the most severe issue and its contamination severity is much more devastating than in other media. These two kinds of OCPs have a higher accumulation level in aquatic organisms. Especially for near-sea organisms, the bio-enrichment effects are relatively

[23] Xie Y. Status analysis on the production and usage of persistent organic pollutants in China [J]. Agriculture and Technology, 2014, 01:250-255.

[24] Yang X. Status Quo of Monitoring Organic Pesticides in China [J]. Environmental Protection of Chemical Industry, 2013, 02:123-128.

serious. In addition, according to Yang Lixin's study,[25] 5 kinds of OCP residues were detected in 19 types of Chinese traditional medicines.

Since the 1980s, PCB production has been forbidden in China. At that point, the total output of PCBs had already reached 10,000 tons, mainly used as additives in paint, printing ink and lubricating oil.[26] According to Ying Xing's survey,[27] eastern China accounted for 45.2% of the total PCB amount, the central region accounted for 35.7%, and the west area was responsible for 19.1%. The survey pinpointed that the contamination of PCBs was not intensively critical in China due to its low yields. However, in the northeast and southern areas, the pollution status gave cause for concern, especially the high content found in river sediments; the PCB concentrations in aquatic organisms also reached dangerous levels. Illegal methods of tackling PCBs and unreasonable choices for locations were largely responsible for these situations.

In China, the main source of PAHs is from human activity and energy utilization processes, including oil and coal burning, oil and petrochemical production, and oil leaking during oil development and transportation. Studies have shown that PAHs in the atmosphere are mainly attached to particulate matters; 70-90% of PAHs were adsorbed on inhalable particulate matter less than 5 microns across; the majority of PAHs were adsorbed on fine particulate matters ($D<1\mu m$).[28] Polycyclic PAHs were the main type found in surface soil. PAH soil pollution was found to be more serious in the north than in the south, and the coastal area was generally better than inland areas.[29] Water pollution also appeared to follow a similar pattern. However, coastal region water pollution conditions varied considerably in different areas.

Dioxin pollution in China is at a moderate level compared to the rest of the world.[30] However, according to published papers,[31,32,33,34] it can be

[25] Yang L, etc. Detection and Determination of 19 Organichlorine Pesticide Residues In Chinese Herbal Medicines by Capillary Gas Chromatography[J]. Chinese Journal of Experimental Traditional Medical Formulae[J]. 2013,04:96-99.

[26] Jiang K, etc. Emissions of Dioxins, Dioxin-like PCBs and PCBs from a Testing PCB Incinerator [J]. Advances in Environmental Science, 1999,01:45-49.

[27] Yang X, etc. A spatial temporal assessment of pollution from PCBs in China. Chemosphere 60 (2005):731–739.

[28] Shi B, etc. Research Review in the Progress of the Regional Geochemistry of Polycyclic aromatic Hydrocarbon in China[J]. Journal of Safety and Environment, 2013, 05:10-20.

[29] Ca Y, et al. Patterns of PAHs concentrations and components in surface soils of main areas in China [J]. Acta Scientiae Circumstantiae, 2012, 32(1): 197-203.

[30] Wu Y, et al. Advances in Pollution Status, Control and Remediation of Soil Contaminated with Dioxins [J]. Soils, 2006, 05:509-516.

detected within the blood and even breast milk in most Chinese residents; these papers commonly pointed out that dioxin levels in the blood were related to residents' ages. [35] According to Man Ren's study, the PCDD/Fs concentration in the Haihe River was higher than in the Pearl River, while the Yangtze River was the least seriously contaminated by dioxin pollution. Nevertheless, the pollution levels of these rivers were still significantly higher than abroad (the Cairo River in Egypt, the Autumn River in Japan, and the Mississippi River in the southern USA). Jingguang Li[36] found, after analyzing 18 provinces in China, that the exposure risk in developed industrial areas was much higher than in other areas; in addition, the pollution in food (dairy products and aquatic products) and pollution in the environment were the key factors that led to exposure risk. Zheng[37] suggested that on the mainland, metal, sodium pentachlorophenol production and solid waste incineration were the main sources of dioxin in China.

3.1. The Distribution Characteristics of POPs in the Chinese Atmosphere

Most substances are exchanged and transferred through the atmosphere, which is also the main environmental component for human livings. In the process of POP production and use, a significant number of POPs are released into the air, and are then spread nationwide via the atmospheric movement.

[31] Deng B, etc. The levels of dioxin in 60 human breast milk samples and the exposure risk factor analysis in Shenzhen [J]. Journal of Preventive Medicine, 2010, 44(3):224-229.

[32] Sun S, etc. Dioxin concentration in human milk in Hebei province in China and Tokyo, Japan: potential dietary risk factors and determination of possible sources. [J]. Chemosphere, 2006, 62(11):1879–1888.

[33] K, Xiao, etc. Polychlorinated dibenzo-p-dioxins and dibenzofurans in blood and breast milk samples from residents of a schistosomiasis area with Na-PCP application in China.[J]. Chemosphere, 2010, 79(7):740–744.

[34] Hedley A J, etc. Breast Milk Dioxins in Hong Kong and Pearl River Delta [J]. Environmental Health Perspectives, 2006, 114(2):202-208.

[35] Hedley A J, etc. Breast Milk Dioxins in Hong Kong and Pearl River Delta [J]. Environmental Health Perspectives, 2006, 114(2):202-208.

[36] Li J, et al. A national survey of polychlorinated dioxins, furans (PCDD/Fs) and dioxin-like polychlorinated biphenyls (dl-PCBs) in human milk in China.[J]. Chemosphere, 2009, 75:1236–1242.

[37] Zheng, GJ, etc. Polychlorinated dibenzo-p-dioxins and dibenzofurans pollution in China: sources, environmental levels and potential human health impacts.[J]. Environment International, 2008, 34(7):1050–1061.

China is facing serious pollutant diffusion problems. In the process of organochlorine pesticide use, 90% of pesticides did not reach the target object, and part of the pesticide residue was trapped in gases or aerosol, which were adsorbed on solid particles in the air and diffused by atmospheric movement.[38] The mixing height of volatile pesticides, such as HCH and HCB, reached nearly ten thousand meters, causing the pollution area to expand continuously. The atmosphere played a crucial role in the process of POP redistribution nationwide.

3.1.1. The Distribution Characteristics of PAHs

The PAH distribution characteristics in the Chinese atmosphere are as follows: (1) the PAH concentration in the air varies considerably between seasons, and generally, winter>autumn>spring>summer; (2) the geographical location has a significant influence on PAH concentration, whereby Southern China is generally lower than Northern China, and coastal areas are lower than inland ones; (3) PAH concentrations in municipalities with high population density are higher than in other less dense regions (urban > rural)[39,40,41] (as shown in Table 3-1).

In accordance with the literature [17, 18, 19, 20], PAHs in the Chinese atmosphere are composed mainly of 4-6 rings, by being attached to air particles.

In addition to the rules above, within the scope of a city, the distribution of PAHs in the atmosphere is also associated with population density, as well as different functional areas of cities. In general, the higher the population density is, the more the PAHs unit concentrates together, thus fostering the higher concentration of PAHs in the local atmosphere (Table 3-2).

---

[38] Xing S, etc. The research progress of Persistent Organic Pollutants in atmosphere[J]. Straits Science, 2-13, 08:29-32.

[39] Zhang J. Distribution Characteristics and Source of Polycyclic Aromatic Hydrocarbons in Various.

[40] Zhou B, etc. Seasonal Variation and Health Risk Assessment of Atmospheric Polycyclic Aromatic Hydrocarbons (PAHs) in the Urban Area of Xi'an [J]. Acta Scientiae Circumstantiae. 2012, 32(9): 2324-2331.

[41] Wang Y, etc. Seasonal Variation and Source Identification of Polycyclic Aromatic Hydrocarbons (PAHs) in Airborne Particulates of Bejing[J]. Environmental Chemistry, 2010, 29(3): 369-375.

Table 3-1. The distribution characteristics of PAHs in main cities of China (ng/m3)

Area	Location	Year	Spring	Summer	Autumn	Winter
Beijingng[42]	Haidian	2014	15.44	10.51	28.78	50.62
Shanghai[43]	Zhabei	2007	42.86	14.91	83.67	54.92
	Jiading	2007	61.733.	24.53	78.07	66.81
Xi'an[44]	Xi'an University of Architecture And Technology	2011	25.9	11.6	37.1	105
Chongqing[45]	Dadukou district	2012	132.92	71.3	107.55	188.8
	Shapingba district	2012	154.47	99.04	130.15	234.12
	Ban'an district	2012	105.58	64.88	90.75	168.91
	Beipei district	2012	36.78	20.27	31.65	66.78
Guangzhou	Liwan district	2003	28.49	17.22	32.13	86.31
	Wushan district		33.96	20.29	31.41	76.82
Nanjing[46]	Dachang	2006	107.26	121.89	111.84	88.6
	Shanxi road		76.276	52.9	42.56	75.82

[42] Wang Q. Study on PM2.5, PAHs and Phthalates in the Air Haze in Haidian of Beijing [D]. 2015, Renmin University of China.

[43] Xiong C. Seasonal Variation and Source Identification of Atmospheric Polycyclic Aromatic Hydrocarbon (PAHs)[D]. Shanghai University, 2008.

[44] Wang Q. Study on PM2.5, PAHs and Phthalates in the Air Haze in Haidian of Beijing [D]. 2015, Renmin University of China.

[45] Tan J, etc. Seasonal Variation of Particulate Polycyclic Aromatic Hydrocarbons Associated with PM10 in Guangzhou, China [J]. Acta Scientiae Circumstantiae, 2005,07:855-862.

[46] Zhang Q, etc. Investigation of Polycyclic Aromatic Hydrocarbons in Atmospheric PM10 in Nanjing City[J]. Environmental Science and Technology, 2007,10:42-44+117-118.

Table 3-2. PAHs in air samples from typical locations of Guangzhou (ng/m^3)[47]

Name	Hotel Lampblack	Guangzhou train station	Southern China botanical garden	Petrochemical company
Bap	6.976	15.840	2.863	1.162
ΣPAHs	427.130	300.130	54.740	19.280

3.1.2. Distribution Characteristics of Organochlorine Pesticides (OCPs)

The pollution level was relatively high in Shenyang and Guangzhou, but low in Beijing and Tianjin. OCP content was detected in the atmosphere in Mount Qomolangma, and although exiguous, it provided direct evidence that atmospheric transmission could result in organic pollution in remote parts of the world (as shown in Table 3-3). Xu Guofei et al.'s[48] study showed that the atmospheric OCP content was characteristically higher in the north than in the south. At the same time, its concentration in air samples in economically developed areas was higher than in other areas.

Table 3-3. The distribution characteristics of OCPs in some areas of China(ng/m^3) [49]

City/Area	Year	HCH	DDT
Shenyang	2009	18.97	42.27
Dalian	2007	0.137	0.064
Beijing	2002	0.24	0.96
Tianjin	1996	0.58	1.87
Hohhot	2007	0.28-1.68	0.04-0.14
Chongming Island	2007	0.38-2.26	0.17-0.98
Guangzhou	2006	109.01	131.14
Mount Qomolangma	2002	0.021	0.005

[47] Zhang A, etc. Content and Characteristics of PAHs and alkanes in Air Pollution Source of Guangzhou City [J]. Journal of Anhui Agriculture Science, 2010, 38(24): 13158-13161.

[48] Zhang Z. Polychlorinated Biphenyls in Chinese Air and Surface Soil: Spatial Distribution Characteristics and Their Inherent Causes [J]. Harbin Institute of Technology, 2010.

[49] Yang X. Status Quo of Monitoring Organic Pesticides in China [J]. Environmental Protection of Chemical Industry, 2013, 02:123-128.

3.1.3. Distribution Characteristics of Polychlorinated Biphenyls (PCBs)

The figure reflects the distribution characteristics of PCBs in the Chinese atmosphere as follows:

1) The PCB content in eastern coastal areas was significantly higher than levels found in inland areas;
2) The PCB level was significantly higher in North China than in South China;
3) The PCB level was significantly higher in urban areas than in rural areas, despite some exceptions wherein the content of individual rural areas was even higher than the city levels. Also, the samples in Qinghai, Tibet and other places showed the presence of PCBs, which proved the fact that persistent organic pollutants in the atmosphere are distributed by means of atmospheric circulation and have long-distance transmission. According to Zhang Zhi's research,[50] the high concentration of PCBs detected in the Chinese atmosphere was mainly concentrated in economically developed, densely populated areas such as the eastern and central regions. The concentration changes of different types of sampling points was not more than 5 times, indicating that the concentration range of PCBs in the Chinese atmosphere was relatively consistent.

3.1.4. The Distribution Characteristics of (PCDD/Fs) of Dioxin

Due to the scarcity of relevant data, the condition of dioxin distribution nationwide cannot be directly determined. However, according to the pollution condition of the 4 districts in Guangzhou, (as shown in Tables 3-4 and 3-5): (1) the concentration distribution regularity: industrial area > residential area > rural area; (2) the dioxin concentrations of low-temperature seasons appeared to be low; (3) there was a correlation between the dioxin concentration and industrial activities; (4) the seasonal distribution characteristics of dioxin: Spring > Winter > Autumn > Summer. Children's respiratory exposure risk

---

[50] Zhang Z. Polychlorinated Biphenyls in Chinese Air and Surface Soil: spatial Distribution Characteristics and Their Inherent Causes [J]. Harbin Institute of Technology, 2010.

and daily intake were significantly raised in the Liwan District, Guangzhou, up to 4.1 I-TEQ/(kg - D), which exceeded the scope of the WHO regulation (1-4 I-TEQ/(kg - D)).[51]

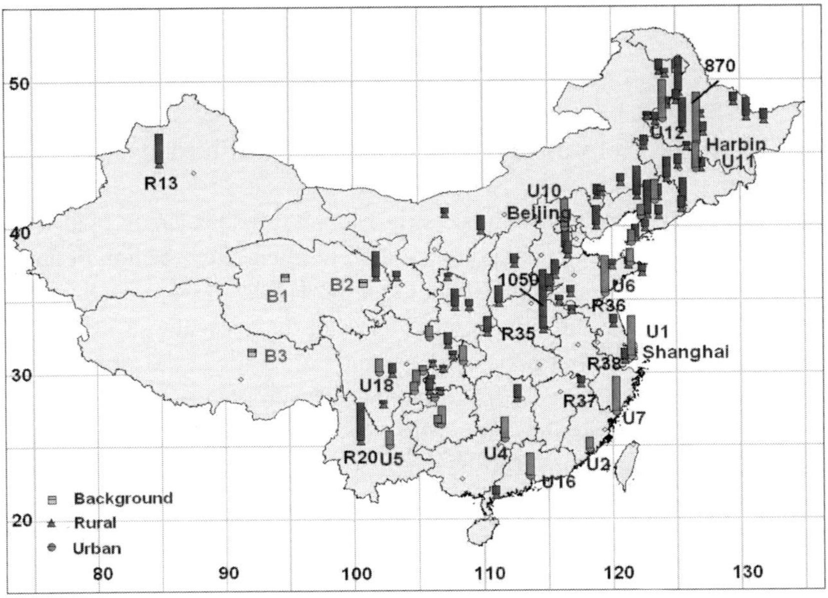

Figure 3-1. The distribution characteristics of PCBs in China.[52]

Table 3-4. The distribution of atmospheric PCDD/Fs in Guangzhou[53]

Guangzhou city	Year	Concentration (fg/m^3)	(fg I-TEQ/m^3)
Huadu district	2007	3815	104.6
Liwan district	2007	12777	430.5
Tianhe district	2007	6963	163.7
Huangpu district	2007	10953	769.3

[51] Yu L. Preliminary Study on Levels of PCDD/Fs in the Atmosphere of Guangzhou and Typical Emission Sources of PCDD/Fs, China [D]. Guangzhou Institute of Geochemistry, Chinese Academy of Sciences, 2007.

[52] Zhang Z. Polychlorinated Biphenyls in Chinese Air and Surface Soil: Spatial Distribution Characteristics and Their Inherent Causes [J]. Harbin Institute of Technology, 2010.

[53] Yu L. Preliminary Study on Levels of PCDD/Fs in Atmosphere of Guangzhou and Typical Emission Sources of PCDD/Fs, China [D]. Guangzhou Institute of Geochemistry, Chinese Academy of Sciences, 2007.

Table 3-5. The seasonal variation of atmospheric PCDD/Fs in Guangzhou (fg I-TEQ/m^3)[54]

Functional area	Spring	Summer	Autumn	Winter
Industrial areas	415	460	1510	337
Other areas	331	402	629	191

Except for the significant seasonal factor, dioxin in Guangzhou's atmosphere also turned out to be geographically sensitive. As shown in Table 3-5, the dioxin content in the industrial areas of Guangzhou accounts for nearly one-fourth of the total amount, especially when Spring comes, when it accounted for approximately 60%, illustrating that the industrial source apportionment of dioxin is significant in Guangzhou. Although it is not beyond the standards in USA and Europe (1000 fg I-TEQ/m^3), such situations should be taken seriously to critically monitor the dioxin exposure risk.

3.2. Distribution Characteristics of POPs in the Water and Bottom Sediments

Although the total quantity of China's water resources is high, a lot of water sources are subject to different degrees of pollution, and persistent organic pollution has become increasingly serious. A variety of POP substances were detected in our seven rivers; in the Pearl River, Haihe River, Huaihe River, and Songhua River, pollution is more serious.[55] Sediment is an important part of the study of water pollution, because persistent organic pollutants in the water mean enrichment in the sediments, which can impact the pollution status of those waters over a long period of time. In addition, when the sediment is in direct contact with the water, pollutants between the two media are constantly exchanged and migrated.

In addition to freshwater resources being put in danger from pollution, in China, the oceanic ecological environment has been seriously threatened. Liu Yan et al. found ultra-trace organic contaminants in China's seawater, especially polycyclic aromatic hydrocarbons, polychlorinated biphenyls and

[54] Su Y, etc. Pollution levels and seasonal variability of atmospheric PCDD/Fs in Guangzhou[J]. Research of Environmental Sciences, 2014, 27(8): 813-819.

[55] Su Y, etc. Pollution levels and seasonal variability of atmospheric PCDD/Fs in Guangzhou[J]. Research of Environmental Sciences, 2014, 27(8): 813-819.

organic chlorine pesticides. Although the concentration is very low, due to the persistence, concentration and long-distance transmission of pollutants, they are still a threat to the marine environment.[56]

Table 3-6. PAH concentration ranges of some watersheds in China[57,58,59,60,61,62]

Watersheds	Sample type	Year	Concentration (ng/L)
Pearl River	river water	2003	15.9-184.2
Qiantang River	river water	2005-2006	70.3-1844.4
Watersheds	Sample type	Year	Concentration (ng/L)
Xiamen Sea	seawater	1999	6960-26900
Yangtze River	river water	2005	242-6235
Tonghui River	river water	2002	192.5-2651

3.2.1. Polycyclic Aromatic Hydrocarbons (PAHs)

Because the polycyclic aromatic hydrocarbons have high hydrophobicity, meaning that they have low solubility in water, the number of rings in water PAHs are mostly low molecular (1-3 rings). Chen Yuyun states in his paper[63] that the Qiantang River water samples with low molecular rings accounted for

[56] Liu Y, etc. The persistent organic pollutants (POPs) enrichment analysis technology in seawater [J]. Environmental Science and Technology, 2010, S1:201-205.

[57] Guo G, etc. Assessment on PAHs Distribution Characteristics and Ecological Risks of Surface water in China [J]. Science China Press, 2012, 05:680-691.

[58] Chen C, etc. Present situation of pollution and ecological effects of polycyclic aromatic hydrocarbons in sediments in China [J]. Liaoning urban and rural environmental science and technology, 2006, 03:57-59.

[59] Ma W, etc. The Distribution Characteristics of Polycyclic Aromatic Hydrocarbons in Soil and Air in China and Numerical Simulation on Large Scale [D]. Harbin Institute of Technology, 2010.

[60] Liu Y, etc. Overview of in-situ Enrichment analytical Technology for Persistent Organic Pollutants (POPs) in Seawater [J]. Environmental Science and Technology, 2010, S1:201-205.

[61] Wang Z. Assessing the ecological risk of Sediment-associated Polychlorinated Biphenyles in Huaihe River [J]. China Environmental Science, 2001, 21(3): 262-265.

[62] Yuan X, etc. Characteristics and Environmental Effects of Polychlorinated Biphenyls in Sediments From Taihu Lake [J]. Resources and Environment in the Yangtze Basin, 2004,13(3): 272-276.

[63] Chen Y. Space-time Distribution, Source and Biological effectiveness of Polycyclic Aromatic Hydrocarbons in Water of Qiantang River [D]. Zhejiang University, 2008.

79.5%. Yang Qingshu[64] found that in the Pearl River, the deeper the vertical depth, the more high-ring PAH molecules (4-6 rings) were found.

According to a comparison of the results based on existing information, the most polluted areas of seawater are the Xiamen waters, while the least polluted is the Pearl River. As can be seen in Table 3-7, the sediments of the Pearl River Delta, South Sea are less contaminated. There is a clear contradiction between the sediment pollution and water pollution in the delta of the Pearl River; the difference between the sampling point and time might be the cause of this result. In addition to prohibiting or reducing the use of POPs in recent years, China has also reduced pollution and made a positive ecological contribution. Also, the distribution of water and sediments is in line with the "North South light weight" of the law; however, based on the ranges from the coastal city, more information is needed on whether it is more than the inland city of pollution do judgment.

3.2.2. Distribution Characteristics of Polychlorinated Biphenyls (PCBs)

China's surface water quality standard was officially executed in 2002 (GB3838-2002 of China), and polychlorinated biphenyls have been listed as a specific project. The standard says that PCB concentrations in uncontaminated water should be lower than 0.1 ng/L. In secondary pollution, the PCB concentration was 50-500 ng/L; 500 ng/L or more was considered severe pollution. However, the pollution status in water and sediments in China is not optimistic; many major rivers have a risk of moderate or severe pollution. Moreover, in China's water, the PCBs mainly exist in the form of low-chlorine substituted monomers, but the high-chlorine PCBs are not volatile, have stable chemical properties, and can be detected in the interstitial water and groundwater.

According to a comparison of the results based on existing information, the most polluted areas of seawater are Xiamen waters, while the least polluted are in the Pearl River. As can be seen in Table 3-8, the sediments of the Pearl River Delta, South Sea are less contaminated. There is a clear contradiction between the sediment pollution and water pollution in the delta of the Pearl

[64] Yang Q, etc. Vertical Distribution, Composition of Polycyclic Aromatic Hydrocarbons in Water Column from Guangzhou Channel in the Pearl River [J]. Marine Science Bulletin, 2008, 06:34-43.

River; the difference between the sampling point and time might be the cause of this result, In addition to prohibit or reduce the use of POPs policies in recent years, China also reduce pollution and make a positive contribution. And the distribution of water and sediments in line with the "North South light weight" of the law, however, ranging from the coastal city needs more information on whether it is more than the inland city of pollution do judgment.

Table 3-7. The PCB concentration ranges of some watersheds in China[65]

Sampling area	Year	Concentration (ng/L)
Tonghui river	2004	32.58-344.9
Jiaozhou bay	1995	9.01-12.63
Huaihe river	1999	0.24-0.28
Yangtze River	2002	0.29-2.00
Pearl river delta	2002	1.1-303.11
Daya bay	2001	91.1-1355.3

Table 3-8. The PCB concentration ranges of sediment from some watersheds in China[66]

Watersheds	Sample type	Year	Concentration (mg/kg)
Qiantang River	river mud	2006	91.3-1835.2
Huangpu River	river mud	2006	107-1707
Jiaozhou Bay	sea mud	2004	20-2200
Yellow River	river mud	2005	464-2621

According to the surface water environmental quality standards (GN3838-2002 of China), Dayawan is likely to be severely polluted (500 ng/L); the Pearl and Tonghui Rivers may appear moderately polluted (50-500 ng/L). On the whole, the PCB distribution in the national water supply is different from other media; the distribution characteristics of PCBs are higher in the south than in the north. In the Pearl River Delta region and the Dayawan region, the

[65] Liu Y, etc. Overview of in-situ Enrichment Analytical Technology for Persistent Organic Pollutants (POPs) in Seawater [J]. Environmental Science and Technology, 2010, S1:201-205.

[66] Liu Y, etc. Overview of in-situ Enrichment Analytical Technology for Persistent Organic Pollutants (POPs) in Seawater [J]. Environmental Science and Technology, 2010, S1:201-205.

PCB content in water was significantly higher than that found in the north of the river. The reason for this result may be that the use of PCBs for industry was more common in the south.

Table 3-9. PCBs concentration ranges in sediments from some watersheds in China[67]

Sampling area	Year	Concentration (ng/L)
Songhua River	2008	0.83-125.53
Haihe River	2006	1.90-81.00
Middle and lower reaches of Yellow River	2005	Tr-5.98
Fujiang-Hangzhou section of the Qiantang River	2005	2.78-16.54
Guanzhou section of the Pearl River	2000	12.88-65.31

In China, in comparison with PAHs, the PCB concentration in sediment was lower, but the water sediments, as shown in Table 3-9, are at risk of being contaminated (10 ng/g), and there is a moderate (50 ng/g) to severe risk of pollution in the Songhua and Pearl Rivers. PCB distribution was roughly the same as water pollution, with the pollution being higher in the south than in the north. According to the paper,[68] in China the primary PCBs in sediments were low chlorine substituents, where 2-5 chlorine substitutes have most.

3.2.3. Organochlorine Pesticides (OCPs)

Organochlorine pesticides have high biological toxicity and are difficult to degrade. At present, in the water monitoring in China, there are the more 666 and DDT. According to China's drinking water standard regulations (CJ3020-93) the concentration of 666 in water is $\leq$1000 ng/L; the concentration of DDT is $\leq$5000 ng/L. Although the data shows that China's water is more than the standard, the increased detectable amount is still worthy of attention.

---

[67] Yuan X, etc. Characteristics and Environmental Effects of Polychlorinated Biphenyls in Sediments From Taihu Lake [J]. Resources and Environment in the Yangtze Basin, 2004,13(3): 272-276.

[68] Yun X, etc. Present Situation and Distribution Characteristics of President Organic Pollutants in Freshwater in China [J]. Environmental Chemistry, 2013, 11: 2072-2081.

Table 3-10. The concentration ranges of PCBs of some watersheds in China (ng/L)[69]

Sampling area	Year	BHC	DDT
Wuhan section of the Yangtze River	2001	0.23-1.90	0.18-4.67
Haihe River mainstream	2007	4.02-25.29	6.88-78.80
Liaohe River	1998	7.59-30.82	Tr-4.16
Yellow River Mengjin wetland	2007	Tr-3.39	Tr.-3.02
Harbin section of the Songhua River	2007	0.002-0.006	Tr
Xijiang River Basin of the Pearl River	2005	1.53-5.86	3.40-13.21

Table 3-11. The PCB concentration in sediment from some watersheds of China (ng/g)[70]

Sampling area	Year	HCH	DDT
Nanjing section of the Yangtze River	2005	Tr	0.336-32.83
Yangtze River estuary	2001	0.54-32.63	0.12-0.57
Haihe River mainstream	2007	2.77-5.9	5.82-45.50
Liaohe River	2009	0.23-1.13	0.01-7.03
Middle and lower reaches of the	2005	0.09-12.88	0.05-5.03
Yellow River Mengjin wetland	2007	Tr	Tr-64.58
Harbin section of the Songhua River	2007	0.0003-0.252	Tr
Pearl River	2002	1.42-3.80	2.22-5.02
Huaihe River	2003	0.88-1.86	1.30-4.19

According to the comparison of the existing data, the pollution of the Haihe Basin is the most serious; Liaohe water is seriously polluted by 666 and less polluted by DDT. On the whole, northern river pollution is more serious than in the south, and Pearl River Basin pollution is more prominent in the south. In addition, in the sediment of the Haihe River Basin, pollution is the most serious, and the pollution of the Songhua River is light. Comparing Tables 3-10 and 3-11, there is a good correlation between water pollution and

[69] Yang X. Status Quo of Monitoring Organic Pesticides in China[J]. Environmental Protection of Chemical Industry, 2013, 02:123-128.

[70] Yun X, etc. Present Situation and Distribution Characteristics of President Organic Pollutants in Freshwater in China [J]. Environmental Chemistry, 2013, 11: 2072-2081.

the bottom sediments, and when comparing the north to the south, pollution is also more serious.

Although in most regions of China, the concentration of OCPs in the sediments is not high, there is a wide-range distribution that can be detected throughout the basin; ultra-trace pesticides do exist and organochlorine pesticides in our country do not show obvious differences between north and south. According to Yuan Xiaoyan[71] [37] et al.'s papers, the interval to resist risk assessment standard effect of DDT is 1.58 ng/g; the effects range median (ERM) was 46.1 ng/g, and according to the standard, in China, part of the sediments are a potential ecological risk.

3.3. Distribution Characteristics of POPs in the Soil

Soil is the growth medium of some biological material and almost all plants. POP pollution in the soil will inevitably lead to the pollutants being absorbed by elementary plants and animals, after which they will constantly transfer to and enrich the food chain (net). Finally, they will cause damage to the environment and do harm to human health. In a 1988 survey, the distribution characteristics of organochloride pesticides in China's soil are: the south>the middle>the north, and the difference between the north and the south is more obvious.

3.3.1. Distribution Characteristics of PAHs

The distribution characteristics of PAHs are reflected in Table 3-12, including:

1) higher concentrations in Beijing, Taiyuan, Shenyang etc., illustrating that the north of China has higher pollution than the south of China;
2) the concentrations in Taiyuan and Nanjing are obviously higher than in Guangzhou and Shanghai, which reflects that inland content is

[71] Yuan X, etc. Characteristics and Environmental Effects of Polychlorinated Biphenyls in Sediments From Taihu Lake [J]. Resources and Environment in the Yangtze Basin, 2004,13(3): 272-276.

higher than eastern coastal areas. Furthermore, there may be data deviation due to fewer sampling points in Shanghai.

Table 3-12. Concentration of PAHs in soil from main cities in China

Region	Number of samples	Year	Concentration (ng/g)
Beijing[72]	233	2008	93.30-13141.46
Shanghai[73]	5	2009	7.5-667.6
Taiyuan[74]	15	2006	980-26230
Shenyang[75]	66	2010	290.9-8492.37
Nanjing[76]	32	2007	312-27580
Guangzhou[77]	74	2007	42-3077
Hong Kong[78]	138	2007	Tr-19500

Table 3-13. Concentration of PAHs in soil from all districts in Tianjin (ng/g)[79]

District	Naikai	Xiqing	Jinnan	Beichen	Dongle	Hangu	Tanggu	Dagang	Jinghai	Jixian
BaP	9.35	118	44.7	38.2	20.6	40.6	42.3	35.4	7.06	20.0
ΣPAHs	360	1490	708	904	699	1360	1170	440	142	382

In addition to the above rules, the distribution of PAHs in soil is related to the population density and the different functional areas within the scope of

72 Peng C. Distribution and Health Risk Assessment of Polycyclic Aromatic Hydrocarbons in Urban Soil of Beijing [D]. Hunan Agricultural University, 2009.

73 Liu Y. Determination, Distribution and Source Apportionment of PAHs in Soils and Sediments in Shanghai, China [J], Tongji University, 2008.

74 Liu F, etc. Pollution Characteristics of Polycyclic Aromatic Hydrocarbons in the Topsoil of Taiyuan City, North China. Earth Science Frontiers, 2008,15(5):155-160.

75 Yang M. Assessment on Health Risk of PAHs from the Soil of the Old industrial District Tiexi, Shenyang[J] Chinese Journal of Ecology, 2013,32(3): 65-681.

76 Research Progress on PAHs Distribution and Risk Assessment of Unban Soil[J]. Chinese Journal of Applied Ecology, 2010, 02:514-522.

77 Research Progress on PAHs Distribution and Risk Assessment of Unban Soil[J]. Chinese Journal of Applied Ecology, 2010, 02:514-522.

78 Research Progress on PAHs Distribution and Risk Assessment of Unban Soil[J]. Chinese Journal of Applied Ecology, 2010, 02:514-522.

79 Zhu Y, Characteristics, Sources Apportionment and Ecological Risks Assessment of Polycyclic Aromatic Hydrocarbons in Soils of Tianjin, China [J]. 2014,33(2):249-255.

the city. On the whole, the greater the population density and the more concentrated the PAHs unit, the higher the PAHs concentration in local soil (Tables 3-12 and 3-13).

3.3.2. Distribution Characteristics of Organochloride Pesticides

The distribution of PCBs in soil has the following characteristics:

(1) The geographical position has a significant effect on the PAHs in the atmosphere. Generally, the southern region is lower than the northern region and the coastal area is lower than the inland area.
(2) The concentration of densely populated areas in the city is higher than in other areas (the city center is greater than the suburbs). According to many papers, the atmospheric polycyclic aromatic hydrocarbons in China mainly exist by adsorbing onto air particulate matter.

Table 3-14. The distribution characteristics of PCBs in main cities in China[80]

Region	Time	HCH	DDT
Pearl River Delta	2000	5.66-22.80	1.51-11.70
Shenyang	2009	2.32-15.90	9.06-111.60
Harbin	2006	0.136-51.76	0.072-28.22
Changbai Mountain	2007	Tr-25.40	Tr-78.70
Guangzhou	2007	0.30-14.90	0.88-697.70
Tongzhou District of Beijing	2007	4.09	26.36
Chengdu	2007	0.01-25.21	0.04-652.98
Guizhou	2007	3.37-8.26	6.45-27.56
Nanjing	2007	2.48-17.80	3.36-74.19
Qingdao	2005	0.41-9.67	3.88-79.55
Chongming Island	2006	0.40-20.00	0.78-163.20
Hohhot	2008	65.64-622.00	66.96-1599.10
Heilongjiang River Valley	2005	1.06	0.87
Yellow River Valley	2009	Tr	63.81-315.65
Lanzhou	2009	0.95-5.94	0.88-108.55

[80] Zhang Z. Polychlorinated Biphenyls in Chinese Air and Surface Soil: Spatial Distribution Characteristics and Their Inherent Causes [J]. Harbin Institute of Technology, 2010.

3.3.3. Distribution Characteristics of Polychlorinated Biphenyls (PCBs)

The distribution of PCBs in the soil, as shown in Figure 3-2, has the following characteristics, which are similar to those of air:

1) Content on the east coast is significantly higher than in inland areas.
2) Urban areas are higher than rural areas, except for some individual rural areas.
3) Northern coastal areas are higher than inland areas, and southern inland areas are higher than coastal areas.

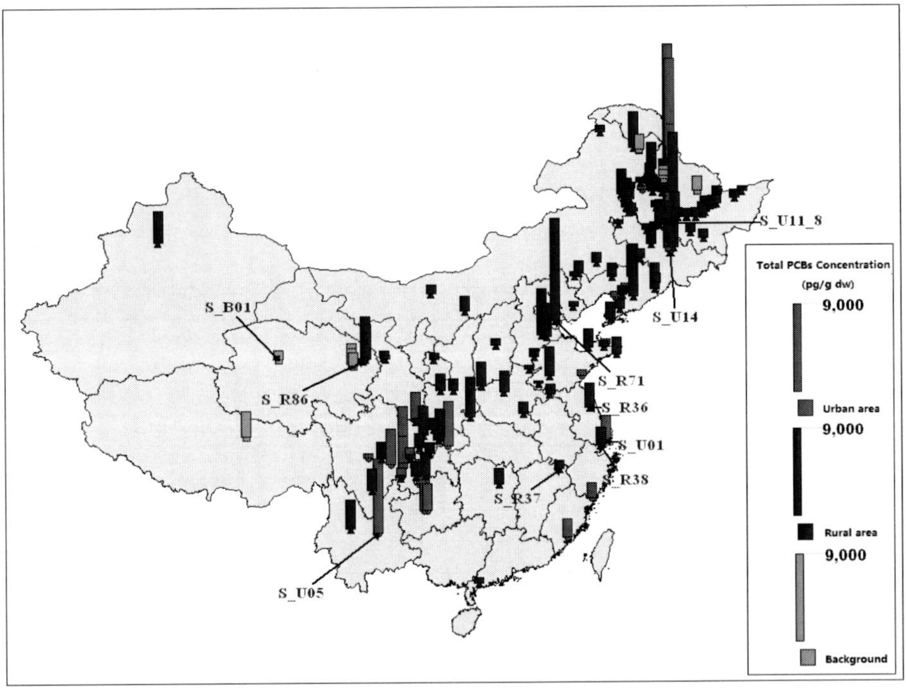

Figure 3-2. The distribution characteristics of PCBs in China.[81]

---

[81] Zhang Z. Polychlorinated Biphenyls in Chinese Air and Surface Soil: Spatial Distribution Characteristics and Their Inherent Causes [J]. Harbin Institute of Technology, 2010.

The existence of PCBs in the background samples of Qinghai, Xizang, etc. can also prove the integrity of the atmosphere. POPs can be spread across a long distance by atmospheric circulation. According to Zhang Zhi's study, the PCBs in the soil mainly come from industrial processes.

3.3.4. Distribution Characteristics of Dioxins PCDD/Fs

The distribution characteristics of dioxins are similar to those of other POPs. Within China, pollution in the north is more serious than in the south, and the coastal regions have more serious pollution than inland areas. However, it is unique to dioxin that its concentration near solid waste incineration plants is obviously higher than in other areas.

Table 3-15. PCDD/Fs concentration in main cities of China

Soil	Area	Number of samples	Year	Concentration (pg/g)
Beijing[82]	Non-contaminated area (Agroforestry)	16	2006	0.2291-3.332
Shanghai[83]	Urban	16	2008	114.24-4037.62
Shanghai[84]	Near the incinerator	12	2008	119.09-3881.44
Dalian[85]	Urban	2	2004	43.3

According to Xue Yuan's research,[86] most PCDDs in the soil are 4-8 chlorine PCDDs, while PCDFs in the soil are mainly 4-8 chlorine PCDFs.

[82] Zhou Z, etc. Preliminary Study on Level and Pattern of PCDD/Fs in Soil in Part of Beijing City [J]. Research of Environmental Sciences, 2006, (6):54-58.

[83] Deng Y, etc. Preliminary Study on the Level of Dioxin-like Compounds in Soil of Shanghai [J]. Journal of Environment and Occupation Medicine, 2008, 25(4):353-355.

[84] Deng Y, etc. Preliminary Study on the Level of Dioxin-like Compounds in Soil of Shanghai [J]. Journal of Environment and Occupation Medicine, 2008, 25(4):353-355.

[85] Yang Z, etc. Research on PCDD/Fs Contents in Several Atmospheric Aerosol and Soil Samples Collected at Shanghai and Dalian[J]. Guangzhou Environmental Sciences, 2004:25-27.

[86] Yang Z, etc. Research on PCDD/Fs Contents in Several Atmospheric Aerosol and Soil Samples Collected at Shanghai and Dalian[J]. Guangzhou Environmental Sciences, 2004:25-27.

Table 3-16. PCDD/Fs concentration in main cities of China (ng/kg) [87]

Number of chlorine	Tetrachlo-ride	Pentachlo-ride	Hexachlo-ride	Heptachlo-ride	Octochlo-ride
Sample 1	14.1	18.0	45.3	189	4718
Sample 2	36.5	Tr	13.3	128	13769
Sample 3	8.6	Tr	20.2	103	1016
Sample 4	12.1	Tr	13.8	127	2259
Sample 5	19.4	Tr	28.1	Tr	7559
Sample 6	21.7	Tr	Tr	73.4	736
Soil background	1.5	Tr	4.7	17.9	282
Ash	16.7	18.0	11.2	10.5	13.4
Fly-ash	7937	5197	4278	1969	652

Table 3-17. The distribution characteristics of PCDFs in soil in Shenyang (ng/kg)[88]

Number of chlorine	Tetrachloride	Pentachloride	Hexachloride	Heptachloride	Octochloride
Sample 1	61.0	12.4	21.1	8.6	<1.0
Sample 2	66.1	22.4	12.8	15.6	<1.0
Sample 3	32.5	7.6	10.9	14.0	<1.0
Sample 4	43.0	16.6	14.4	18.0	<1.0
Sample 5	43.6	13.1	15.6	54.0	<1.0
Sample 6	48.4	28.3	21.1	13.7	<1.0
Soil background	5.5	2.3	2.9	2.3	<1.0
Ash	342	227	134	67.0	<1.0
Fly-ash	51089	23431	7812	3500	<1.0

[87] Xue Y, etc. Preliminary Study on Distribution of Organochlorine Pesticides and Poly chlorinated Biphenyls in Surface Soils and Atmosphere Along the Xihe River in Shenyang City [J]. Rock and Mineral Analysis, 2011, 30(1):27-32.

[88] Xue Y, etc. Preliminary Study on Distribution of Organochlorine Pesticides and Poly chlorinated Biphenyls in Surface Soils and Atmosphere Along the Xihe River in Shenyang City [J]. Rock and Mineral Analysis, 2011, 30(1):27-32.

3.4. Characteristics of Persistent Organic Pollutant Distribution in Biological Bodies

Research about POPs in organisms mostly focuses on aquatic organisms. In China, it's universally known that POPs exist in aquatic organisms, as PAHs, OCPs, PCBs and PCDD/Fs have all been detected in aquatic plants and animals. In addition, when aquatic plants and dairy products are served as food, the pollution status of POPs in those food items attracts a great deal of attention. According to numerous examples in the literature, TCDD can be detected in the blood and even human milk of people from many areas in China, and all the literature points to a correlation between blood TCDD concentration and the age of the residents. Furthermore, the concentration of dioxin-like chemicals is usually higher in those who frequently eat dairy products and seafood.

3.4.1. OCP Concentration in Aquatic Organisms

It can be seen from the table that the pollution level of the Pearl River far exceeds that of other water bodies, which may be relevant to the overuse of OCPs in the Pearl River Delta in the 1980s. In addition, overall, the residual concentration of OCPs in organisms is higher in China, and in some areas, has reached quite a serious degree.

Table 3-18. OCP concentration in aquatic organisms from some watersheds in China (ng/g) [89]

Aquatic organisms	Year	HCH	DDT
Yangtze River	2005	2.3-38.7	4-286.83
Xiamen	2005	0.66-1.98	6.25-46.90
Lianyungang	2006	Tr	6.4
Quanzhou Bay	2004	0.26-0.90	0.74-64.00
Pearl River	2002	21.97-3925.87	1981.74-42490.78
Shenzhen Bay	2004	0.08-1.85	1.84-286.83
Yuexi Sea	2007	0.84-14.00	0.60-18.40

[89] Yang L, etc. Detection and Determination of 19 Organichlorine Pesticide Residues In Chinese Herbal Medicines by Capillary Gas Chromatography [J]. Chinese Journal of Experimental Traditional Medical Formulae [J]. 2013,04:96-99.

3.4.2. PAHs in Organisms

It can be seen from the table above that the concentration of PAHs in organisms is even higher in vegetables than it is in meat, which reflects the fact that China should put more measures in place to control the addition of PAHs in food processing; it also shows that there is a gap in food standards between mainland China and Hong Kong.

Table 3-19. PAH concentration in some organisms from different cities of China[90]

Organism	Place	Year	Concentration (mg/kg)
moss	South mountain	2005	310-1340
meat and fish	Hong Kong	2004	15.5-118
vegetable	Tianjin	2004	2250-7820
leaf	Beijing	2003	69.3

3.4.3. PCBs in Aquatic Organisms

It can be seen from the table that the pollution of the Pearl River Delta is the most serious, while the concentration of PCBs in the aquatic organisms of the Bohai Sea is the lowest.

The result may be related to the water quality and industry circumstances near the monitoring point.

In addition, the table above also reflects the fact that southern watersheds' PCBs pollution in aquatic organisms is more serious than the pollution in northern watersheds, which is similar to the water pollution comparisons mentioned previously.

[90] Xie Y. Status analysis on the production and usage of persistent organic pollutants in China [J]. Agriculture and Technology, 2014, 01:250-255.

3.4.4. Concentration of PCDD/Fs in Human Milk

The main cause of the difference in TCDD and PCBs toxicity concentrations is the different degree of industrialization and different eating habits. In developed regions, such as Shanghai, a high degree of industrialization and a higher proportion of fatty foods both caused the higher toxic concentration; in some predominantly agricultural areas, such as Ningxia, there may be a correlation between low toxic concentrations in human milk and low degrees of industrialization.

In addition, given China's rapid industrialization and urbanization, the difference in the degree of pollution between urban and rural areas is becoming smaller and smaller. PCB pollution levels in urban and rural areas have become closer, but the human milk concentrations of TCDD in urban areas is higher than in rural areas.

Table 3-20. PCBs in aquatic organisms from some watersheds of China[91,92]

Place	Year	Species	Concentration (ng/g)
Bohai Bay	1996	Fish	5.4
Dalian Bay	2002	Fish	44.3
Jiaozhou Bay	1996	Fish	273.3
Yangtze River delta	2004	Fish	60.69
Pearl River delta	2003	Fish	635.7
Songhua River	1989	Shellfish	61.3
Huaihe	2001	Shellfish	5.96

[91] Xie Y. Status analysis on the production and usage of persistent organic pollutants in China [J]. Agriculture and Technology, 2014, 01:250-255.

[92] Yang X, etc. A spatial temporal assessment of pollution from PCBs in China. Chemosphere 60 (2005):731–739.

Table 3-21. Concentrations of PCDD/Fs in human milk from different areas of China[93,94,95,96,97]

Province		Lipid content g/100ml	TEQ-PCBs pg/g lipid	TEQ-PCDD/Fs pg/g lipid	TEQ-PCDD/Fs pg/g lipid	Dairy intake pgTEQ/kg.bw/d
Heilongjiang	rural	3.5	1.01	1.88	2.89	14.2
	urban	3.5	1.11	1.82	2.93	14.4
Liaoning	rural	4	2.68	5.5	8.18	45.8
	urban	3.3	1.49	2.8	4.29	19.8
Hebei	rural	2.6	1.88	4.24	6.12	22.3
	urban	3	1.14	5.57	6.71	28.2
Shanxi	rural	4	0.93	1.66	2.59	14.5
	urban	3.9	1.18	3.99	5.17	28.2
Ningxia	rural	4.2	0.61	2.52	3.13	18.4
	urban	4.5	0.82	2.49	3.31	20.9
Henan	rural	5	1.19	3.17	4.36	30.5
	urban	2.8	1.68	4.44	6.12	24
Shanghai	rural	3.4	2.99	6.67	9.66	46
	urban	3.5	3.36	6.65	9.92	48.6

---

93 Deng B, etc. The levels of dioxin in 60 human breast milk samples and the exposure risk factor analysis in Shenzhen [J]. Journal of Preventive Medicine, 2010, 44(3):224-229.

94 Sun S, etc. Dioxin concentration in human milk in Hebei province in China and Tokyo, Japan: potential dietary risk factors and determination of possible sources. [J]. Chemosphere, 2006, 62(11):1879–1888.

95 K, Xiao, etc. Polychlorinated dibenzo-p-dioxins and dibenzofurans in blood and breast milk samples from residents of a schistosomiasis area with Na-PCP application in China. [J]. Chemosphere, 2010, 79(7):740–744.

96 Hedley A J, etc. Breast Milk Dioxins in Hong Kong and Pearl River Delta[J]. Environmental Health Perspectives, 2006, 114(2):202-208.

97 Li J, etc. A national survey of polychlorinated dioxins, furans (PCDD/Fs) and dioxin-like polychlorinated biphenyls (dl-PCBs) in human milk in China.[J]. Chemosphere, 2009, 75:1236–1242.

Province		Lipid content g/100ml	TEQ-PCBs pg/g lipid	TEQ-PCDD/Fs pg/g lipid	TEQ-PCDD/Fs pg/g lipid	Dairy intake pgTEQ/kg.bw/d
Fujian	rural	3.8	2.32	2.1	4.42	23.5
	urban	3.9	1.33	3.05	4.38	23.9
Jiangxi	rural	3	2.61	4.35	6.96	29.2
	urban	4.1	1.76	5.21	6.97	40
Hebei	rural	4.9	1.61	3.07	4.68	32.1
	urban	3	2.11	4.58	6.69	28.1
Sichuan	rural	4.4	1.24	2.65	3.89	24
	urban	2.7	1.72	4.17	5.89	22.3
Guangxi	Rural	5.1	1.58	3.46	5.04	36
	urban	5	2.09	3.36	5.45	38.2

Chapter 4

DEMANDING ANALYSIS OF LEGAL REGULATIONS OF PERSISTENT ORGANIC POLLUTANTS IN CHINA

The pollution of persistent organic pollutants has become a major environmental issue of common concern for the international community, and determining how to effectively control POPs has become an important task of the government. The international community has established the basic legal framework for dealing with POPs, and governments from all countries have expressed their response to the policy and their position on dealing with POPs. The international obligation to deal with POPs and policies to deal with them from major countries has put external pressure on China. At present, in China, the problem of persistent organic pollutants harming human health and the ecological environment is increasingly serious. The rapid development of industry and the social economy has resulted in a significant growth in demand for chemicals, which is why China now faces tremendous pressure to control the pollution caused by POPs. The basic national conditions of a large population, a low level of economic development, an unreasonable industrial structure, and a fragile ecological environment make POP pollution control a major and urgent task in China. The law is one of the most effective ways to solve environmental problems, and while the prevention and control of POPs pollution can be based on external pressure and internal demand, the current relevant laws and regulations have some obvious defects. Therefore, in China, the further strengthening and perfecting of the laws for the prevention and control of POPs pollution is of the utmost urgency and necessity.

4.1. External Pressure for Legislation of POPs

According to the existing international legal documents that deal with POPs pollution, China should bear the obligation. The attitudes and policies regarding POPs pollution from America, Japan, the EU and other countries will provide a reference for China, and will also create external pressure on China. Fulfill relevant international treaty obligations and the political attitude from major countries to deal with the POPs pollution which are composed to external pressure for China.

4.1.1. Pressure regarding the International Responsibility for the Prevention and Control of POPs Pollution

In 1972 the "United Nations Declaration on the Human Environment" was passed in the "United Nations Conference on the Human Environment", which marked the point at which the international community began to be concerned about environmental issues. Then, in 1992 the "Rio Declaration on Environment and Development" was passed during the "United Nations Conference on the Human Environment and Development" which was a milestone in human society in terms of solving environmental problems through cooperation. The responsibilities and obligations for the environmental protection of all countries were further strengthened in the "Declaration on the Human Environment." In addition, this declaration put forward the concept of "sustainable development" and "environmental management". The two conventions stated the general requirements for POPs pollution control for each contracting party in order to protect human health and the ecological environment. However, the "Basel Convention on the control of transboundary movements of hazardous wastes and their disposal", (thereafter referred to as the "Basel Convention") stated that certain hazardous chemicals and pesticides in the process of international trade must follow the prior informed consent procedure of the Rotterdam Convention (thereafter referred to as the "Rotterdam Convention" or "the PIC Convention") to the "Stockholm Convention on persistent organic pollutants" (thereafter referred to as the "Stockholm Convention" or "POPs Convention"). All of these actions reflect the fact that the international community attaches great importance to hazardous waste, hazardous chemicals and pesticides, and the prevention and management of POPs pollution. The three international treaties constitute the international legal framework for POPs prevention and management, which gave firm directions for parties who deal with the pollution of POPs and presented the standard requirements. International pressure and power is needed to build POPs legislation for China, and China needs to bear the responsibility within the international legal framework.

1. "Basel Convention" and the States' Responsibility

The United Nations Environment Program held the Environmental Protection Conference and passed the "Basel Convention" in Switzerland on March 22, 1989. The Basel Convention regulations were implemented on September 22, 1995, and in the same year, the "Amendment to the Basel Convention" was passed in Geneva.[98] The Convention is composed of the preface and the body, which includes 29 clauses and 6 components. Its purpose is to ban the transboundary transfer of hazardous wastes, and it deals with pollution in other countries, especially in developing countries, to prevent such international transfer. 99 The 2005 Amendment prohibits developed countries from exporting hazardous wastes to developing countries, and the provisions state that developed countries must stop the exportation of hazardous waste recycling to developing countries before the end of 1997.

China joined the Convention on September 4, 1991. In accordance with the provisions of the Convention, the contracting parties not only enjoy the rights of prohibition of hazardous wastes and other wastes, but they also bear the following responsibilities: each Party must take measures to ensure that hazardous wastes and other domestic wastes are limited to the lowest possible level; they must ensure the provision of adequate facilities for the disposal of hazardous wastes and other environmental wastes to maintain effective management; they must ensure the minimization of the cross-border transfer of wastes; and the exportation of waste to waste importing countries, especially to developing countries, is legally prohibited. In addition, the Convention also provides strict licensing conditions for the waste transfer and prohibits illegal transportation, which is considered a crime.

2. "Rotterdam Convention" and the States' Responsibility

The Food and Agriculture Organization of the United Nations and the United Nations Environment Program formulated the "Rotterdam Convention"

---

[98] http://baike.baidu.com/view/392900.htm?fromtitle=%E5%B7%B4%E5%A1%9E%E5%B0%94%E5%85%AC%E7%BA%A6&fromid=1700535&type=syn.

[99] Han Depei, a course in the law of environmental protection (the sixth edition) [M]. Law Press in China, 2012 (3):470.

in Rotterdam on September 10, 1998, which entered into force on February 24, 2004. The Convention is composed of the preface and the body, which includes 30 clauses and 5 annexes. The purpose of this convention is to protect human health from potentially harmful effects, including the health of consumers and workers, and also to protect the environment from certain hazardous chemicals and pesticides in international trade.

The Convention went into effect in China on June 20, 2005. The scope of this Convention included the prohibition or strictly limited use of chemicals, especially dangerous pesticides. According to the provisions of the Convention, States' parties are required to implement the decision procedure for the import and export of some very dangerous chemicals and pesticides, namely the prior informed consent procedure (PIC). That is to say, the importing countries have the right to obtain relevant data about banning or strictly limiting the use of chemicals from other countries, and can decide whether to agree, restrict, or prohibit future chemical importation into the country, and share this decision with the exporting country. The exporter will share this decision with the domestic export sector so that they can make arrangements and ensure that the international transportation of goods is not in violation of the importing country that is deciding the case. The goal of PICs is to communicate related information about the characteristics of certain hazardous chemicals in international trade, to develop a procedure for the import and export of these types of chemicals, and to inform relevant parties about these decisions. In this way, the shared responsibility and cooperation of parties can be promoted in the international trade of chemicals, thereby it could protect human health and the environment from the harm caused by such chemicals and promote the use of environmentally sound manner.

3. "Stockholm Convention" and the States' Responsibility

The "POPs Convention" was passed in Stockholm in Sweden on May 23, 2001, and entered into force on May 17, 2004. Currently, there are 124 members, including China. The Convention is composed of the preface and the body, including 30 clauses and 5 annexes, and it is the first convention stating that 12 Persistent Organic Pollutants need to be controlled.[100] Its purpose is to reduce or eliminate the emission and release of POPs, to prevent adverse effects in all stages of their life cycle, and to stimulate the

[100] These 12 kinds of POPs including 10 POPs which were intended to produce, namely, aldrin, dieldrin, endrin, toxaphene, seven chlorine, mirex, DDT (DDT), six chlorobenzene and PCBs pesticides and industrial chemicals; two no by-products, namely dioxins and furans.

development and utilization of alternative processes and chemicals to avoid harm to human health and the environment from POPs.[101]

In accordance with the provisions of the Convention, state parties need to perform the following main obligations: to prohibit and eliminate the production of all kinds of POPs (including pesticides and industrial chemicals); to strictly control environmental pollution and health risks in the process of production and the use of POPs; to reduce or eliminate unintended byproducts as POPs (Dioxin); research, development and monitoring; publicity and education; to make a national implementation plan for controlling POPs; to report to the conference of the parties and the establishment of a performance evaluation mechanism.[102] The Convention is considered another important measure through which to further strengthen the management and control of toxic chemicals after the "Basel Convention" and the "Rotterdam Convention", all of which are considered milestones in the international community.

4.1.2. The Pressure from POPs Prevention Policy and the Position of the Main Countries

The development of the global chemical industry, on one hand, brings convenience to human production and life, but on the other hand, it causes great danger to human health and the ecological environment. All countries, especially the major developed countries, must not only contribute to building and enforcing an international treaty, but must also make domestic policies to ensure the safety and effective control of these chemicals.

The outstanding performance in Europe in this regard is due to regulations passed in the European Parliament and Council on 2006/12/18; these have been officially adopted and came into effect on 2007/6/1. This referred to the "Regulation concerning the Registration, Evaluation, Authorization and Restriction of Chemicals (to be called the REACH regulations)." The law established the new management system to implement the registration, evaluation, and charter for a market including 30,000 kinds of produced or imported existing chemicals, and to implement strict restrictions to eliminate

---

[101] Han Depei, a course in the law of environmental protection (the sixth edition) [M]. Law Press in China, 2012 (3):471.

[102] Liu Janguo, etc. The Guideline of Laws and Regulations on China's POPs Management [M]. The China environment press, 2013(7):1.

the persistent, bioaccumulative, and carcinogenic chemicals that have high environmental and health risks. [103] Following the introduction of the EU REACH regulations, Japan followed suit, continuing to issue strict laws and regulations concerning chemical management. [104] In 2010, the Japanese government required enterprises engaged in the production of chemicals to report the import and the use of chemicals to the government once each year, with the purpose of strictly managing the chemicals that may cause environmental and health hazards; this system is called "the Japanese version of the REACH."[105] Due to the implementation of the EU REACH regulations, there was a problem facing American chemicals as to whether the REACH regulations were to be implemented in America.[106] In October of 2009, the Obama administration announced a reform of the National Chemicals Control Act - "the Toxic Substances Control Act."

New measures for chemical regulation put forward by the EU, Japan and the USA have fully demonstrated those countries' strengthened determination to regulate the production, use, and sale of chemical products. In particular, the implementation of REACH in the European Union and Japan, and improved importation chemical standards, the "green trade barrier" will mean at least 730 kinds of exports to the EU will face harsh registration, evaluation, authorization and other chemical tests.

4.2. Domestic Demand of POPs Prevention Legislation

After the Industrial Revolution, human society began to face large-scale environmental problems. The root cause of environmental problems is the lack of understanding and respect for nature. People try to override the natural world, with themselves as the leaders, in accordance with their own will to dominate all things in nature. However, there is a serious contradiction between the unlimited growth of human needs and the limited supply of our

[103] Commission of the European Communities. White Paper on strategy for a future chemicals policy [R]. Brussels:[s.n.], 2001.

[104] Hongbiao Bian, Japan's Step by EU REACH Regulation [J]. WTO Economic Herald, 2009 (2):86.

[105] Hongbiao Bian, Japan's Step by EU REACH Regulation [J]. WTO Economic Herald, 2009 (2):86-87.

[106] Peng Yue, There are some obstacles in the "USA REACH" introduction [J], household electrical appliances, 2008 (8):39-40.

planet's natural environment. With the intensification of this contradiction, the environmental problem is becoming increasingly serious. Urbanization and industrialization will make the POPs pollution situation even worse, and the Chinese government is facing enormous pressure to prevent and control POPs pollution. To control the generation of POPs pollution, China has undertaken active policies and actions, and has also gained certain achievements. Policies and goals in response to POPs pollution have determined the direction for coping with POPs pollution legislation.

4.2.1. The Basic Facts of POPs Pollution in China

In China, industry and agricultural processes produce a large number of POPs every day. Such pollution poses a serious threat to human health, and also leads to the destruction of the ecosystem. In agricultural production, large-scale use of pesticides leads to contaminated crops, causing a series of food safety issues, while also destroying the soil and polluting groundwater. In the process of industrial production, to save on operating costs, heavy polluting enterprises do not use clean production technology, and indiscriminately discharge toxic and harmful substances containing POPs. The release of POPs-containing dust into the atmosphere causes air pollution. The release of POPs-containing illegal sewage into rivers destroys water resources. The reckless disposal and indiscriminate dumping of POPs-containing solid waste only creates more and more POPs-contaminated sites. Other industries, such as catering, will generate a lot of garbage in China; this is often treated by incineration or sent to a landfill, which also causes serious environmental pollution problems.

4.2.2. Challenges Facing the Prevention of POPs

iChina is a highly populous country; the current population is close to 1.4 billion. China currently faces high population density, high resource consumption, unbalanced regional development, a wealth gap and various other issues, all of which are affecting the level of environmental protection in China. A lot of POPs-containing waste is produced in the process of resource consumption, which greatly exceeds the carrying capacity of the environment. As a developing country, China still needs to develop its economy, and in the process, the demand for resources will continue to increase the emissions of pollutants, including POPs, which will further aggravate environmental pollution and exacerbate its effects. Obviously, China is facing severe challenges, considering that history tells us that in exchange for economic development, the environment will pay a heavy price. If attention is not paid to environmental problems, this environmentally damaging behavior will not be contained, and China's future development will not be sustainable.

4.2.3. China's Policies and Actions for the Prevention of POPs Pollution

The Chinese government fulfilled the "Stockholm Convention", and in July 2007, officially launched the implementation of "The People's Republic of China National Implementation Plan for the Stockholm Convention on Persistent Organic Pollutants" (referred to as the "National Implementation Plan"). According to the actual situation in the country phases, sub-regional, sub-sector to carry out compliance activities. The "National Implementation Plan" mainly includes building institutions, policies and regulations to reduce or eliminate emissions from the intentional production and use of POPs, to reduce and eliminate emissions of unintentionally produced POPs, to reduce emissions from POPs stockpiles, to identify POPs-contaminated sites and to promote sound environmental management.[107] The Ministry of Environmental Protection organized member departments to implement this plan, and cooperated closely with other relevant departments to promote the implementation work.[108]

[107] Liu Jianguo, etc. The Guideline of Laws and Regulations on China's POPs Management [M], China environment press, 2013 (7):2.

[108] Liu Jianguo, etc. The Guideline of Laws and Regulations on China's POPs Management [M], China environment press, 2013 (7):3-5.

Chapter 5

ANALYSIS OF THE LEGAL STATUS OF THE PREVENTION AND CONTROL OF POPS POLLUTION

The pressure that China faces at home and abroad regarding POPs pollution prevention and control is the external impetus for legislation on POPs pollution prevention and control. Also, the insufficiency of existing legislation for POPs pollution prevention and control is the internal cause of POPs pollution prevention legislation. In order to perfect the relevant laws for China's POPs pollution prevention and control, a status quo analysis of China’s POPs pollution control legislation must be conducted.

At present, in China, the legal provisions on POPs stipulated from a macro perspective fall within legal fields, such as the Environmental Protection Law, Industry Law, Pollution Prevention and Control Law, and the Occupational Safety and Public Health Law.

In addition, there are also specific regulations and department rules concerning the management and disposal of pesticides and industrial chemical POPs, dioxin-like POPs pollution control, POPs waste, and polluted site management, covering fields such as environmental protection, the green industry, pollution control, occupational safety and public health, POPs pesticides and industrial chemical management, POPs pollution emission control and its waste and contaminated sites management, and many other aspects.

5.1. At the Macro Level

5.1.1. The Environmental Protection Law

POPs pollution is the environmental problem caused by the emission of POPs into the atmosphere, water, and soil as a result of the production and living processes of human beings. Measures can be taken to protect and improve the environment by inhibiting POP emissions and maintaining the stability of the ecosystem.

1. The Relevant Provisions of the Environmental Protection Law of the People's Republic of China 2014 (hereinafter referred to as the Environmental Protection Law 2014)

As the fundamental legislation in the environmental protection field, the Environmental Protection Law 1989 was amended on April 22, 2014, and entered into force on January 1, 2015. The Environmental Protection Law 2014, amended from the Environmental Protection Law 1989, has been referred to as the most rigid law in environmental legislation history. It was developed for the purposes of protecting and improving environment, preventing and controlling pollution and other public nuisances, safeguarding public health, promoting ecological civilization, and enhancing sustainable economic and social development. The aim of the Environmental Protection Law 2014 being to protect the ecosystem and human health is consistent with the legislative purpose of POPs pollution prevention and control legislation. Environmental protection is our nation's basic policy, which is applicable to every field of environmental protection. The fourth chapter of the Environmental Protection Law 2014, which is *Prevention and Control of Pollution and Other Public Nuisances,* stipulates that: the production, storage, transportation, sales, use and disposal of chemicals and other radioactive substances shall abide by the relevant provisions of the state to prevent pollution of the environment (article 48). As toxic pollutants which are significantly detrimental to human health and the ecosystem, POPs belong to the category of chemicals or pollutants requiring control, applicable to this clause.

2. The Relevant Provisions of Clean Production Laws and Regulations

In China, a soft law to encourage enterprises to carry out clean production, "Cleaner Production Promotion Law of the People's Republic of China" (hereinafter referred to as the Cleaner Production Promotion Law) came into force in January 2003. It plays an important role in pollution prevention and risk control in the production and use of poisonous and harmful chemicals, including POPs. Article 19 of the Cleaner Production Promotion Law stipulates that enterprises shall replace highly toxic or highly harmful raw materials with non-toxic or harmless raw materials or low-toxicity or low-harm raw materials. Article 28 of the Cleaner Production Promotion Law also makes provisions for a Cleaner Production Audit which provides that the use or discharge of poisonous and harmful material in the production of enterprises shall warrant a regular Cleaner Production Audit; the rules of the Cleaner Production Audit shall be made by the relevant departments for economic and trade together with the administrative department of environmental protection of the State Council. In addition, the Interim Measures for the Clean Production Audit passed in August 2004 and the implementation of the "provisions on the key enterprises of Cleaner Production Audit procedure" in December 2005 further detailed the requirements for the Cleaner Production Audit of enterprises. Interim Measures for the Clean Production Audit made relevant provisions relating to the mandatory requirements regarding the scope, procedure and organization management of the Clean Production Audit, and the list of toxic and harmful substances was required to be more stringently audited. The auditing provisions of the Cleaner Production Promotion Law lay the foundations for the production, use and emissions of POPs by relevant companies and for the replacement of POPs with other substances to eliminate or minimize pollution control.[109]

3. The Relevant Provisions of the Environmental Impact Assessment Law and Regulations

The Environmental Impact Assessment Law of the People's Republic of China (hereinafter referred to as the Environmental Impact Assessment Law) which entered into force in September 2003 made provisions regarding all kinds of economic behavior including the POPs pollution control of hazardous chemicals. According to the Environmental Impact Assessment Law, the analysis, prediction and assessment of environmental impact of planning and construction projects, and countermeasures to prevent or mitigate the possible adverse environmental impacts, as well as the monitoring and tracking of the planning and the construction projects could have a precautionary effect on the

[109] Liu Janguo, etc. The Guideline of Laws and Regulations on China's POPs Management [M]. The China environment press, 2013(7):13.

pollution of poisonous and harmful materials and risk control. In August 2009, the implementation of the planning environmental impact assessment regulation made relevant provisions regarding the environmental impact of all kinds of development and construction planning that may result in POPs emission and pollution, as well as scientific requirements and environmental pollution prevention measures.

4. The Regulation on the Administration of Recycling and Disposal of Waste Electric and Electronic Products

The Regulation on the Administration of Recycling and Disposal of Waste Electric and Electronic Products was implemented, making specific provisions regarding control of the production and application of electric and electronic products, the application and identification of non-toxic or low toxic and harmful substances, and the harmless disposal and management of waste products, as well as the phase-out of some techniques and the prohibition of certain production processes.

5. Trial Measures for the Environmental Information Disclosure

Trial Measures for the Environmental Information Disclosure which came into effect in May 2008 aimed to promote and standardize the disclosure of environmental information by the competent authorities for environmental protection and enterprises, to protect the rights and interests of the citizens, legal persons and other organizations regarding environmental information, and to promote public participation in environmental protection. In China, against the backdrop of a lack of effective measures for the control of toxic chemical pollutants such as POPs, the Trial Measures will aid in the control of POPs pollution emissions.

5.1.2. The Green Industry Legislation

In January 2009 the Circular Economic Promotion Law of the People's Republic of China (hereinafter referred to as the Circular Economic Law) was implemented; this was a typical green industry legislation that aimed to promote the reduction, reuse, and recycling of resources and waste from the production processes. The Circular Economic Law made provisions regarding the adjustment of the structure of chemical industries, including those industries that produce POPs. It introduced rules regarding the use and recycling of resources as well as waste pollution prevention and risk control during the production process, including the production and use of toxic and harmful substances, the harmless disposal of hazardous waste, and the elimination of harmful chemicals; it also provided incentives for environmentally friendly alternatives.

5.1.3. The Pollution Prevention and Control Legislation

In China, the POPs pollution prevention and control provisions mainly exist among the contents of the Air Pollution Prevention and Control Law of the People's Republic of China (hereinafter referred to as the Air Pollution Prevention and Control Law), the Water Pollution Prevention and Control Law of the People's Republic of China (hereinafter referred to as the Water Pollution Prevention and Control Law) and the Solid Waste Pollution Prevention and Control Law of the People's Republic of China (hereinafter referred to as the Solid Waste Pollution Prevention and Control Law).

The Air Pollution Prevention and Control Law, which came into effect in September 2000, made provisions regarding limits on the emissions of toxic atmospheric pollutants, and the emergency prevention of the leakage of poisonous and harmful gas caused by atmospheric pollution accidents. In addition to the provisions of the corresponding toxic pollutant emission limits and accident emergency prevention system, the Water Pollution Prevention and Control Law, which came into effect in June 2008, also provided the discharge permit system and the phase-out of outdated technology, processes and equipment that could cause serious water pollution. In 2013 the revised the Solid Waste Pollution Prevention and Control Law provided the basic rules for the harmless management and disposal of POPs, including: a hazardous waste list and unified system for the identification and classification standards (Article 51), an identification system for hazardous waste containers and packaging (Article 52), a transfer system for hazardous waste (Article 59) a qualification system for the collection, storage, utilization and disposal of hazardous waste (Article 57), and the emergency management of environmental safety and accidents (Articles 62, 64).[110]

The three laws mentioned above are binding in terms of air, water and waste pollution prevention and control regarding POPs emissions.

[110] Liu Janguo, etc. The Guideline of Laws and Regulations on China's POPs Management [M]. The China environment press, 2013(7):18.

5.1.4. Occupational Safety Legislation

Because POPs present a hazard to the human health of the personnel engaged in the economic activities of dangerous chemicals, the revised Work Safety Law of the People's Republic of China made provisions for the evaluation of the production and management of dangerous chemicals and the safety of hazardous waste disposal, as well as the protection of the rights of supervision, and the management and emergency rescue of major hazards. These provisions are conducive to the control of occupational safety and health risks. In addition, the Prevention and Control of Occupational Diseases Law of the People's Republic of China (hereinafter referred to as the "Law on the Prevention and Control of Occupational Diseases), implemented in 2011, provided a legal basis for the protection of the occupational health of laborers. The law stipulates the responsibilities for the prevention of occupational diseases caused by poisonous and harmful materials, as well as the prohibition, restriction and elimination of materials that cause serious occupational hazards, and so on.

5.1.5. Public Health Law

China has an old saying, "the food to Ann first," and food security relates to people's personal safety and healthy lives. The frequency of food safety accidents in recent years has promoted the legislation and implementation of the Food Safety Law of the People's Republic of China in 2009.

The mandatory enforcement of the food safety standards stipulated in the law and the ban on the production of food not meeting the food quality standard, have to some extent prevented the production of POPs food with poisonous and harmful materials, which will be assist in protecting the public from the POPs risk.

5.2. At the Micro Level

In China, in addition to management laws and regulations regarding POPs at the macro level, there are also micro-level rules for the management of pesticides and industrial chemical category POPs, as well as the control of dioxins emissions, waste and pollution site management regulations and other measures and guidelines.

5.2.1. The Related Laws, Regulations and Administrative Rules About the Pesticides-Category Pops Management

In China, the pesticides category POPs regulations and department rules include the pesticide management ordinance (2001), the Agricultural Product Quality Safety Law (2006), pesticide production management measures (2004), the measures for the safe use of pesticides sets (1982), the ban on the use of pesticides (2004), the environmental management regulation on the first-time import of chemicals and the import and export of toxic chemicals (2007) and the notice about the ban on the production, circulation, use and import and export of DDT, chlordane, mirex, and hexachlorobenzene (2009). These regulations and department rules constitute a legal safety system regarding the registration, production, use and import and export of pesticides. The systems for pesticide registration and pesticide production, license review, import and export management of pesticides, the pesticide industry, quality and waste disposal and supervision and management constitute the basic legal system of pesticide management.

5.2.2. The Industrial Chemical Category POPs Management Regulations, and Administrative Rules

In China, industrial chemical category POPs related regulations and department rules include the Hazardous Chemical Materials Safety Management Regulations (2011), the Measures for the Implementation of the Dangerous Chemicals Production Enterprise Security License (2011), the Measures for the Safety Supervision and Management of Dangerous Chemical Construction Projects (2012), the Measures for the Management of Dangerous Chemicals Business License (2012), the Environmental Management Measures for New Chemicals (2010), the Environmental Management Measures on the import of chemicals for the first time and the import and export of toxic chemicals (2007), the Notice on Strengthening the Management of the import and export registration of Poisonous Chemicals (2010) and the Restricted List of toxic chemicals to be imported and exported in China (2012). These regulations and department rules constitute the legal system for the management of industrial chemicals and safety regarding industrial chemical registration, production, use, transportation, storage, import and export. In addition, to limit the import and use of the polychlorinated biphenyls (PCBs) power plants, regulations and department rules have been enacted specifically for PCB management, such as Containing PCBs (PCBs) Power Plant Operation Management Regulations (1991), and the regulation on Preventing the Polychlorinated Biphenyls (PCBs) Power Equipment and its Waste from Polluting the Environment (1991).

5.2.3. The Relevant Normative Documents on the Dioxin-Like POPs Pollution Emission Control

In October 2010 the Ministry of Environmental Protection, the Ministry of Foreign Affairs, the National Development and Reform Commission, the Ministry of Science and Technology, the Ministry of Industry and Information Technology, the Ministry of Housing and Urban-rural Development, and the Ministry of Commerce and State Administration of Quality Supervision, Inspection and Quarantine jointly issued the Guidelines on Strengthening the Dioxin Pollution Prevention (hereinafter referred to as the Dioxin Guidance), and put forward clear goals and specific measures for the control of dioxins pollution over the following five years. The Dioxin Guidance aims to control the growth trend of dioxin emissions; its contents are mainly concerned with the optimization and upgrading of industrial structures, the control of dioxin pollution in key industries, the establishment of long-term mechanisms, and technology research, development and demonstration for dioxin pollution prevention. In addition, the Chinese government formulated a series of standards and technical specifications on dioxin pollution control.

5.2.4. The Relevant Normative Documents on the Waste Management of POPs and Contaminated Site Management

In accordance with the provisions of the Stockholm Convention on Persistent Organic Pollutants, POPs waste belong to the category of hazardous waste management. In China, Solid Waste Pollution Prevention and Control (revised in 2013) is the particular legislation relevant to solid waste management; this law makes provision for the harmless management of POPs waste. Also, a relatively complete national environmental management system for hazardous waste was established following a series of regulations and standards regarding the import, management, discharge, transfer and disposal of POPs, together with the Solid Waste Pollution Prevention and Control.[111]

[111] Liu Janguo, etc. The Guideline of Laws and Regulations on China's POPs Management [M]. The China environment press, 2013 (7):22.

Chapter 6

DEFECTS IN THE LEGAL GOVERNANCE OF POPs IN CHINA AND SUGGESTIONS FOR IMPROVEMENT

In accordance with the Stockholm Convention, the Chinese government has not only continuously studied new technologies to reduce the generation and emission of POPs, but has also committed to solving the environmental problems caused by POPs through legal means. However, China started later in addressing this problem of the legal governance of POPs, so inevitably, there are some defects in the realms of education, legislation, technology, law enforcement, justice, and environmental protection. However, the Environmental Law, revised in April 2014, brought new opportunities for the Chinese legal governance of POPs. It further improved the relevant laws, regulations and standards, strengthened law enforcement, promoted judicial reform, and enhanced awareness of the need for environmental protection and the relevant laws, which have been the priorities in terms of China's response to the problem of POPs.

6.1. DEFECTS IN THE CHINESE LEGAL GOVERNANCE OF POPs

The Chinese government has repeatedly stressed that China should not repeat the same mistakes as developed countries; however, in the past thirty years, environmental pollution incidents have occurred frequently in China. The state of the environment has been deteriorating; people's local environments and the whole ecological environment are both facing a huge threat. China's rapid economic growth has come at great environmental cost, following the path of "treatment after pollution." To reverse the current situation of serious environmental problems it necessary to rely on strong legal measures.

China has legislated in many fields related to POPs, but either the legal system for the prevention and control of POPs or its specific contents is imperfect. There are some problems, including the lag in enforcing legislative principles, the incomplete legal system and the inadequate response system etc. In addition, there are also some deficiencies in administrative law enforcement and judicial and law-abiding consciousness.

6.1.1. The Lag in Legislative Ideas

China has a lot of legal documents relating to POPs, but compared with western countries, there is obviously a lag in legislative ideas.

The legislative idea contains the legal value and spirit; its rationality and scientific nature affect the structure and the effectiveness of the legal system. The legislative idea and the legal purpose are closely related to the law principle; the legal purpose is actually the embodiment of the legislative idea, and the legal principle is the implementation of the legal purpose and legislative idea. Therefore, in order to examine the legislative idea for the prevention and control of POPs, we should start from the legal purposes and principles of the law.

The principle of coordinated development is a fundamental principle of environmental law. The principle requires the overall planning of environmental protection, economic and social development in order to achieve common economic, social and environmental benefits. However, some laws relating to POPs do not reflect the coordinated development of environment, economy and society. For example, the "Law of the People's Republic of China on the Prevention and Control of Environmental Pollution by Solid Waste" Article 1 was enacted for the purpose of preventing and controlling environmental pollution by solid waste, safeguarding human health and promoting the development of the socialist modernization drive. According to the provision, the purpose of the legislation is not only to prevent and control environmental pollution but also to promote sustainable economic and social development. But when environmental protection conflicts with economic growth, which is more important? The problem cannot be solved by applying the provisions. Obviously, this example does not comply with the legislative purpose of the coordinated development of the environment, economy and society.

6.1.2. The Legal System Is Not Perfect

The legal system is a unified system formed by a country's existing legal norms and departments. Legal norms are the basic units of the legal system. As a new area of law, the prevention and control of POPs should be composed of a comprehensive law, special separate regulations for certain types of POPs and other legal documents related to POPs.

From the perspective of the current Chinese legal system for the prevention and control of POPs, the biggest problem is the lack of a comprehensive law. The author believes that, due to the seriousness of POPs, it is sufficient to become a type of contamination with an emphasis on air pollution, water pollution and solid waste pollution. It is necessary to enact a comprehensive law for the prevention and control of POPs. Currently, there are legislative precedents around the world specifically designed for the prevention and control of POPs. Regulation (EC) No 850/2004 of the European Parliament and the Council of 29 April 2004 on persistent organic pollutants and the amending Directive 79/117/EEC is by far the most exhaustive specialized regulation for the problem of POPs worldwide.[112] The directive would be very valuable for China in legislating for the control of POPs.

In addition, other relevant laws with different emphases regarding legislative purposes do not focus on the prevention and control of POPs, so it can not play a very good treatment effect for POPs. Currently, laws involving some aspects of the control of POPs are scattered among the "Law on the Prevention and Control of Atmospheric Pollution," "Law on the Prevention and Control of Water Pollution" and "Law on the Prevention and Control of Environmental Pollution by Solid Waste." These three laws are focused on air pollution, water pollution and solid waste pollution, respectively, and their main purpose is to solve the environmental problems caused by air pollution, water pollution and solid waste pollution, rather than to legislate for the

[112] Zhou Bingyan, etc. Oversea Environmentally Sound Management of Persistent Organic Pollutant Wastes [J], Environmental Protection of Chemical Industry, 2006 (5):430.

prevention and control of POPs specifically. Therefore, the relevant provisions are inadequate and vague, and generally insufficient for the effective prevention and control of POPs.

The absence of the "Law on the Prevention and Control of Soil Pollution" is an important factor in POPs pollution. In China, POPs constitute one of the culprits responsible for soil pollution, especially in rural areas. Heavy use of substandard pesticides and insecticides in agricultural land has resulted in the POPs contamination of a large area of arable land each year, but so far China has no specific legislation for soil pollution.

Other specific regulations related to the management of POPs are also lacking to date. There is no special regulation for POPs contaminated sites in China. Existing laws and regulations for the management of contaminated sites are scattered among the "Law on the Prevention and Control of Environmental Pollution by Solid Waste," "Measures on environmental management of hazardous chemical waste," "Interim Provisions on Administration of Environmental Protection in Cases of Waste Imports" and other laws and department regulations. These provisions are simply a matter of principle; they lack the requirements for specific measures to be taken, and are therefore difficult to implement. Also, there is nothing from the legislative perspective regarding the management of contaminated soils and contaminated sites. This weakens the concept of soil protection for enterprise managers.[113] In addition, with respect to Europe and other developed countries, China's research into dioxin pollution control is nearly 20 years behind. Specific legislation regarding dioxins has yet to be introduced. Currently only individual standards and norms regarding dioxin emissions have been developed, and these standards and norms simply remain under the direction for the incineration of waste. Control of dioxins in other related industries, such as pollutants, is essentially lacking.[114]

[113] Xue Nandong, etc.: POPs risk control and environmental repair of contaminated site [M], Science Press, 2011 (12):196.

[114] Zhang Xiaoling, Process and Measure of Dioxin Emission Reduction in Developed Countries [J]. Environmental and Ecology in the Three Gorges, 2013 (1):36.

6.1.3. Laws and Regulations Related to POPs Pollution Control Are Not Perfect

Regarding the content, most of the laws and regulations related to the prevention and control of POPs are provisions in principle. There are some problems: the operability and enforceability are insufficient; the responsibility is unclear; management is not effective, and so on. For example, the "Regulation on management of old electrical and electronic products" implemented on May 1, 2013 appears beneficial, but the actual operability is not strong; in fact, it has failed to control the POPs pollution caused by the recycling of Waste Electronic & Electric Equipment (WEEE).[115] Guiyu in Shantou City, Guangdong Province, known as "the world's largest electronic garbage dump," is China's underground "Holy Land" of the e-waste industry, as the location for the annual dismantling of as much as one million tons of e-waste. This e-waste has caused very serious pollution of both water and soil during the recycling process. Hundreds of people are suffering from cancer caused by environmental pollution every year; this place is well known as a "cancer village" of China. In accordance with the basic principle of environmental law, i.e. "polluter pays," the ecological damage and personal injury to the local residents should have been the responsibility of the relevant companies. However, due to the inoperability of the law and the low cost of breaking the law, coupled with the concept of only GDP misleading, no improvement has been made to the long-term pollution situation in this area.

Another example is that the law on solid waste pollution prevention and control makes some stipulations regarding the collection, repositioning, use, disposition of solid waste for individuals and units; it also regulated the harmless disposition of recycled waste. These clauses serve only to offer the rules regarding contaminated sites in principle, but no implementation details exist to clarify the management responsibility, investment mechanism, or prevention and control measures for the contaminated sites. In particular, the closed, merged or transferred enterprises have left behind toxic and hazardous wastes and contaminated sites, but their treatment and disposal incurs huge expense. Unfortunately, the principle of cost-sharing is not clear. There is a lack of appropriate legal accountability mechanisms regarding the existing

---

[115] LiuYi, WEEE Recycling Environmental Impacts and Management Strategies [J], Environmental Science and Management, 2012 (3):13.

regulations on POPs management and pollution control.[116] In addition, the punishment for violating the regulations concerning POPs contaminated sites is inadequate, and is therefore ineffective as a legal deterrent.

6.1.4. Defects of POPs Pollution Prevention in Enforcement, Judicial and Law-Abiding Processes

First of all, in China the main problem is the capacity for environmental administration and enforcement is low. For example, there is a lack of comprehensive, specific legislation regarding the prevention and control of POPs pollution, although there are provisions pertaining to it scattered within environmental basic law, environmental specific law, environmental regulations, departmental regulations and other normative legal documents. The delineation of authority between administrative departments is still unclear regarding the division of power and the existence of an overlap between departments gives the impression of a fragmented system, beset by buck-passing and competing interests. China's environmental management departments are subject to central and local governments at all levels; they are not financially independent; they compete in the pursuit of economic interests and the only GDP ideas' misleading. Local governments intervene in the law enforcement activities of the environmental authorities and the relevant administrative law enforcement departments. In addition, because of local protectionism, cross-regional POPs pollution problems cannot be solved effectively. For example, in order to pursue their particular economic interests, some areas condone or even cover up the illegal sewage disposal of local enterprises and other damage to the environment, ignoring environmental issues. In Guiyu Guangdong the e-waste category POPs pollution has caused serious consequences, and the government's omission is an important factor.

Secondly, POPs pollution in China has caused extensive ecological damage and personal injury to residents. In attempting to solve these problems through judicial means, it appears that filing, trial and judgment are very difficult to enforce. In terms of the environmental torts caused by POPs, who constitutes the plaintiff? In the event of a hearing, how can a causal relationship be determined? When dealing with pollution incidents across

[116] Xue Nandong, etc.: POPs risk control and environmental repair of contaminated site [M], Science Press, 2011(12):196.

different regions, which court is authorized to accept the case? What are the problems of the causal relationship between the fact of damage and the behavior? When the defendant is unable to compensate the plaintiff, what is the solution? A series of problems exists in the judicial practice of POPs pollution which has plagued the court system.

Finally, other factors include opaque government information, citizens' participation rights not being guaranteed, government inaction, the low cost of breaking the law and the high cost of being law-abiding. These factors have contributed to low levels of law-abiding behavior from enterprises. In addition, the citizens themselves do not pay enough attention to environmental and legal education. Government propaganda is insufficient, and citizens lack environmental and legal consciousness, so that the relationship between the public, enterprises and the government has become extremely fragile due to the frequent occurrence of habitat destruction, and large-scale mass incidents.

6.2. Suggestions for the Legal Governance of POPs Pollution Prevention in China

To sum up, in order to achieve effective results in the legal governance of POPs pollution, China has to work on its legislation, law enforcement, judicature and law-abidingness one by one. With this in mind, we put forward some suggestions:

6.2.1. Change the Legislative Idea

Due to the current situation of serious environmental pollution in China, there is a need to change the legislative ideas in the field of environmental protection. Fortunately, in 2014 April the newly revised "Environmental Protection Law" brought an opportunity to change the legislative idea of POPs pollution control. The newly revised environmental protection law's general provisions in Article 4 provide that: the state shall adopt economic and technological policies and measures favorable for the conservation and circulatory use of resources, protection and improvement of environment and harmony between human and nature, so as to coordinate economic and social development with the work of environmental protection. This provision laid the foundation and direction for the transformation of the legislative idea of

POPs pollution control. The environmental legislation idea was a priority of economic development, so China's environment and people have paid a heavy price. China needs to follow the legislative spirit of modern environmental law, to seek a balance between environmental protection and socioeconomic development, make environmental benefits, economic benefits and social benefits of unity; this is the inevitable direction for China's POPs pollution legislation.

6.2.2. Establish a Sound Legal System for POPs Management

The problems with China's current legislation for POPs pollution prevention and control include the legislation idea lag, the imperfect legal system and insufficient relevant laws and regulations and so on. This is largely because of the lack of a specific and comprehensive POPs Pollution Prevention Law.

Because the range of POPs pollution control is too broad, it is very difficult in the short term to revise and improve relevant legislation, but it is possible to introduce a comprehensive law for POPs pollution prevention and control in a relatively short period of time. This could play a leading role in the legal governance of POPs pollution. POPs Pollution Prevention Law needs only to make some provisions in principle in order to guide separate relevant regulations for their development, revision and implementation, and related special regulations are needed to fulfill specific operational requirements. Therefore, a comprehensive POPs Pollution Prevention Law should be developed as soon as possible, as well as a timely revision of the relevant laws and regulations; this is the inevitable approach to achieve adequate legislation for the control of POPs pollution.

Given the current situation of serious soil pollution, the introduction of a “Soil Pollution Prevention Law” is urgent. It is essential in the effective prevention and treatment of soil pollution due to POPs. In addition, specific rules and regulations for the control of dioxin pollution and POPs contaminated sites should also be developed as soon as possible.

6.2.3. Improve the Relevant Legal Provisions for Preventing and Controlling POPs Pollution

It is necessary to perfect the mechanism for sharing the responsibility of pollution. For recycling products containing POPs, producer responsibility should be extended to the entire life cycle of product, making the manufacturer responsible for product recovery, recycling and final disposal. Individual enterprises are unable to bear the costs of management of the disposal of toxic and hazardous wastes and contaminated sites. Also, this kind of pollution is very difficult to trace in terms of responsibility; therefore, we must provide a mechanism for joint funding among polluters, beneficiaries and all levels of government departments in the law, to provide financial security for the treatment of POPs contaminated sites.

In addition, in view of the existing problem of the low penalty for breaking the law and the high cost of compliance, stricter punishment methods should be prescribed by law such as the “daily penalty” in order to increase the degree of punishment. This should be used to make up for the lack of funds to control POPs pollution, creating an effective legal deterrent.

6.2.4. Suggestions Regarding Law Enforcement, Judiciary and Legal Compliance for Preventing POPs Pollution

In administrative law enforcement, first of all, the government departments at all levels and administrative law enforcement departments should change the priority from traditional economic development, adhering to green GDP ideas to realize the sustainable development of the economic and social environment, as economic and social development cannot be achieved at the expense of the environment. Secondly, a clear division should be made between the functions and accountability of various departments, distinct from the administrative law enforcement authority, in order to facilitate efficient administrative enforcement. Finally, environmental liability with lifelong responsibility should be established; that is to say, the chief executive from the government at all levels must assume lifelong responsibility for local environmental protection.

In the judicial process, we should explore the establishment of special courts to deal with environmental tort cases, including environmental protection cases involving a wide range of areas. In the court system we need professional judges for determining causality; if the judge does not have a solid theoretical knowledge and experience of environmental law, the court will not be able to establish a mechanism for internal expert consultation, and many environmental cases will not be conducted as fair trials. In addition, for cross-regional POPs environmental pollution cases, we should follow the example of western countries in setting up an environmental protection circuit court to avoid the issue of the court jurisdiction not being clear. In 2013 the amendments to the "Civil Procedure Law" clearly defined the "civil environmental public litigation system." In 2014 the newly revised "environmental protection law" further clarified the system. However, in China now, the progress of environmental public interest litigation is not smooth, the scope of plaintiff qualification is too narrow, and many environmental pollution cases result in a finding that the local government did not do anything. There are no provisions for public interest litigation regarding the environmental administration in the "administrative litigation law," which can mean that relevant environmental cases cannot be filed, so we should further explore and develop the system of public interest environmental litigation.

In terms of compliance with the law, the first priority is that all information affecting the environment should be fair, and transparent to the public, and there should be a guarantee of the right of the stakeholder to know information affecting the environment, which would enable the public to actively participate in environmental protection. At the same time, the government should increase environmental protection and legal propaganda, and strengthen citizens' environmental protection and legal consciousness. For enterprises, environmental protection and environmental law lectures could be organized to minimize damage to the environment in the production process, making enterprises consciously abide by the law. Citizens should be persuaded to actively participate in various environmental protection activities, starting from scratch and continuing to cultivate awareness of the law regarding environmental protection.

Law is a weighing to a country. Governments must pay attention to their legal role in solving environmental problems which affect human survival and the earth's ecological system. POPs pose a great threat to the living environment of mankind and the whole ecosystem, necessitating not only solutions to the problems with innovative clean technology, but also the use of legal weapons to control such pollution. POPs pollution is basically caused by people, and the law is needed to regulate harmful human behavior; therefore, we need to pay more attention to the legal governance of POPs pollution.

CONCLUSION

With China's rapid economic development some serious environmental problems have also arisen, including problems of water pollution, solid waste pollution, and POPs pollution, which are increasingly grim situations. Due to the explosive growth trends of environmental pollution incidents in recent years, the government is also aware that the extensive mode of economic development in China has come at the expense of the environment, and President Xi Jinping is determined to deal with pollution. He has gained some achievements in the past two years, but there is a long way to go for Chinese environmental protection. The control of POPs pollution has also reached crunch time; the legal governance mechanism is the inevitable method for controlling POPs.

References

Ajay Kumar Goel. Anthrax: A disease of biowarfare and public health importance [J]. *World Journal of Clinical Cases,* 2015(1).

Allan Astrup Jense. Why has the Stockholm Convention Banned the Brominated Flame Retardant Hexabromocy Clododecane (HBCDD)[C]. 2014-9-21.

Boening D W. *Environ. Safety,* 1998, 39:155-163.

Ca Y, etc. Patterns of PAHs concentrations and components in surface soils of main areas in China [J]. Acta Scientiae Circumstantiae, 2012, 32(1): 197-203.

Chen C, etc. Present situation of pollution and ecological effects of polycyclic aromatic hydrocarbons in sediments in China [J]. *Liaoning urban and rural environmental science and technology,* 2006,03:57-59.

Chen Y. Space-time Distribution, Source and Biological effectiveness of Polycyclic Aromatic Hydrocarbons in Water of Qiantang River [D]. Zhejiang University, 2008.

Commission of the European Communities. White Paper on strategy for a future chemicals policy [R]. Brussels:[s.n.],2001.

Deng B, etc. The levels of dioxin in 60 human breast milk samples and the exposure risk factor analysis in Shenzhen [J]. *Journal of Preventive Medicine,* 2010, 44(3):224-229.

Deng Y, etc. Preliminary Study on the Level of Dioxin-like Compounds in Soil of Shanghai [J]. *Journal of Environment and Occupation Medicine,* 2008, 25(4):353-355.

Frignani M, Bellucci L G, Carraro C et al. *Chemosphere,* 2004, 54: 1563-1572.

Gai Nan. Ecological geochemistry study of organochlorine pesticides and polychlorinated biphenyls of Ruoergai wetland [D]. *China Academy of Geological Sciences Press,* 2012-06-08.

Guo G, etc. Assessment on PAHs Distribution Characteristics and Ecological Risks of Surface Water in China[J]. *Science China Press,* 2012,05:680-691.

Han Depei, A course in the law of environmental protection (the sixth edition) [M]. *Law Press in China,* 2012(3):470.

He Shichang. Chemical characteristics of organochlorine pesticides in Antarctic route atmosphere [D]. Third Institute of Oceanography of State Oceanic Administration, 2013.

Hedley A J, et al. Breast Milk Dioxins in Hong Kong and Pearl River Delta [J]. *Environmental Health Perspectives*, 2006, 114(2):202-208.

Hong Bing. The study of POPs in Chinese alligators [D]. *Anhui Normal University Press,* 2014.

Hong H, etc. Seasonal variation of PM10-bound PAHs in the atmosphere of Xiamen, China [J]. *Atmospheric Research*, 2007, 85: 429-441.

Hongbiao Bian, Japan's Step by EU REACH Regulation[J*]. WTO Economic Herald,* 2009(2):86-87.

Hou Deshen. Study on the distribution characteristics and ecological effects of persistent toxic substances in the western region of Inner Mongolia[D]. *Mongolian University Press,* 2014.

http://baike.baidu.com/view/392900.htm?fromtitle=%E5%B7%B4%E5%A1%9E%E5%B0%94%E5%85%AC%E7%BA%A6&fromid=1700535&type=syn.

Hu Riledawa. Distribution characteristics of OCPs, PCBs and heavy metal in soil, grass and park in Mongolia [D]. *Mongolian University Press,* 2014.

Huijuan Xiu. Environmental toxicology [M]. Beijing: *Environment Science Press of China,* 2003.

Jiang K, etc. Emissions of Dioxins, Dioxin-like PCBs and PCBs from a Testing PCB Incinerator [J]. *Advances in Environmental Science*, 1999, 01:45-49.

Jones K C, de Voogt P. *Environ. Pollut.,* 1999, 100: 209-221.

Li J, etc. A national survey of polychlorinated dioxins, furans (PCDD/Fs) and dioxin-like polychlorinated biphenyls (dl-PCBs) in human milk in China. [J]. *Chemosphere,* 2009, 75:1236–1242.

Liu F, etc. Pollution Characteristics of Polycyclic Aromatic Hydrocarbons in the Topsoil of Taiyuan City, North China. *Earth Science Frontiers,* 2008,15(5):155-160.

Liu Janguo, etc. The Guideline of Laws and Regulations on China's POPs Management [M]. T*he China environment press*, 2013(7):1.

Liu Ling, Chui Guangbo, Hao Zhenchun. *Journal of hydraulic engineering,* 2003, 6:6-13.

Liu Wenmin. Effect of PBDEs on expression of genes related to the immunity of zebrafish [D]. *Shandong Normal University Press,* 2014-03-26.

Liu Y, etc. Overview of in-situ Enrichment analytical Technology for Persistent Organic Pollutants (POPs) in Seawater [J]. *Environmental Science and Technology,* 2010,S1:201-205.

Liu Y, etc. The persistent organic pollutants (POPs) enrichment analysis technology in seawater [J]. *Environmental Science and Technology,* 2010,S1:201-205.

Liu Y. Determination, Distribution and Source Apportionment of PAHs in Soils and Sediments in Shanghai, China [J], Tongji University, 2008.

LiuYi, WEEE Recycling Environmental Impacts and Management Strategies [J], *Environmental Science and Management,* 2012(3):13.

Ma W, etc. The Distribution Characteristics of Polycyclic Aromatic Hydrocarbons in Soil and Air in China and Numerical Simulation on Large Scale [D]. *Harbin Institute of Technology,* 2010.

Peng C. Distribution and Health Risk Assessment of Polycyclic Aromatic Hydrocarbons in Urban Soil of Beijing[D]. Hunan Agricultural University, 2009.

Peng Yue, There are some obstacles in the "USA REACH" introduction [J], household electrical appliances, 2008(8):39-40.

Research Progress on PAHs Distribution and Risk Assessment of Unban Soil[J]. *Chinese Journal of Applied Ecology,* 2010, 02:514-522.

Rui Chen, Qing Yan Xu, Bai Cheng Liu. Simulation of the Dendrite Morphology and Microsegregation in Solidification of Al–Cu–Mg Aluminum Alloys [J]. *Acta Metallurgica Sinica,* 2015(2).

Shi B, etc. Research Review in the Progress of the Regional Geochemistry of Polycyclic aromatic Hydrocarbon in China [J]. *Journal of Safety and Environment,* 2013, 05:10-20.

Su Y, etc. Pollution levels and seasonal variability of atmospheric PCDD/Fs in Guangzhou [J]. *Research of Environmental Sciences,* 2014, 27(8): 813-819.

Sun S, etc. Dioxin concentration in human milk in Hebei province in China and Tokyo, Japan: potential dietary risk factors and determination of possible sources.[J]. *Chemosphere,* 2006, 62(11):1879–1888.

Tan J, etc. Seasonal Variation of Particulate Polycyclic Aromatic Hydrocarbons Associated with PM_{10} in Guangzhou, China [J]. *Acta Scientiae Circumstantiae,* 2005,07:855-862.

ten Tusscher G W, Koppe J G. *Chemosphere,* 2004, 54: 1329-1336.

Wang Chunhui, Wu Shaohua, Zhou Shenglu, et al. Research progress on the characteristics and environmental behavior of typical soil POPs [J]. *Environmental Chemistry,* 2014-11-15.

Wang Q. Study on $PM_{2.5}$, PAHs and Phthalates in the Air Haze in Haidian of Beijing [D].2015, Renmin University of China.

Wang Y, etc. Seasonal Variation and Source Identification of Polycyclic Aromatic Hydrocarbons (PAHs) in Airborne Particulates of Bejing [J]. *Environmental Chemistry,* 2010, 29(3): 369-375.

Wang Z. Assessing the ecological risk of Sediment-associated Polychlorinated Biphenyles in Huaihe River [J]. *China Environmental Science,* 2001, 21(3): 262-265.

Wen Sheng, Hui Yang, Yang Fangxing. Polychlorinated dibenzo-p-dioxins (PCDDs) and dibenzofurans (PCDFs) in surface sediment and bivalve from the Changjiang Estuary, China [J]. *Chinese Journal of Oceanology and Limnology*, 2008-03-15.

Wu Y, etc. Advances in Pollution Status, Control and Remediation of Soil Comtaminated with Dioxins [J]. *Soils,* 2006, 05:509-516.

Xiao K., etc. Polychlorinated dibenzo-p-dioxins and dibenzofurans in blood and breast milk samples from residents of a schistosomiasis area with Na-PCP application in China. [J]. *Chemosphere,* 2010, 79(7):740–744.

Xie Y. Status analysis on the production and usage of persistent organic pollutants in China [J]. *Agriculture and Technology,* 2014, 01:250-255.

Xing S, etc. The research progress of Persistent Organic Pollutants in atmosphere [J]. *Straits Science,* 2-13, 08:29-32.

Xiong C. Seasonal Variation and Source Identification of Atmospheric Polycyclic Aromatic Hydrocarbon (PAHs)[D]. Shanghai University, 2008.

Xu G, etc. Research Progress on Organochloride Pesticides in Atmosphere in China [J], *Journal of Anhui Agriculture Science,* 2009, 37(6):2690-2693.

Xue Nandong, etc: POPs risk control and environmental repair of contaminated site [M], *Science Press,* 2011(12):196.

Xue Y, etc. Preliminary Study on Distribution of Organochlorine Pesticides and Poly chlorinated Biphenyls in Surface Soils and Atmosphere Along the Xihe River in Shenyang City[J]. *Rock and Mineral Analysis*, 2011, 30(1):27-32.

Yajun An, Fang Huang. A Review of Mg Isotope Analytical Methods by MC-ICP-MS[J]. *Journal of Earth Science,* 2014-10-15.

Yang L, etc. Detection and Determination of 19 Organichlorine Pesticide Residues In Chinese Herbal Medicines by Capillary Gas Chromatography [J]. *Chinese Journal of Experimental Traditional Medical Formulae [J].* 2013,04:96-99.

Yang M. Assessment on Health Risk of PAHs from the Soil of the Old industrial District Tiexi, Shenyang [J] *Chinese Journal of Ecology,* 2013,32(3): 675-681.

Yang Q, etc. Vertical Distribution, Composition of Polycyclic Aromatic Hydrocarbons in Water Column from Guangzhou Channel in the Pearl River [J]. *Marine Science Bulletin,* 2008,06:34-43.

Yang Qiaoyun. Study on the exposure and health effects of typical organic pollutants in the body [D]. Peking University, 2014-06-01.

Yang X, etc. A spatial temporal assessment of pollution from PCBs in China. *Chemosphere* 60 (2005):731–739.

Yang X. Status Quo of Monitoring Organic Pesticides in China [J]. *Environmental Protection of Chemical Industry*, 2013, 02:123-128.

Yang Z, etc. Research on PCDD/Fs Contents in Several Atmospheric Aerosol and Soil Samples Collected at Shanghai and Dalian[J]. *Guangzhou Environmental Sciences,* 2004:25-27.

Yu Gang. Persistent organic pollutants: a new global environmental issues[M]. *Science Press,* 2005.

Yu Gang. Persistent organic pollutants: a new global environmental issues[M]. Beijing: *Science Press,* 2005.

Yu L. Preliminary Study on Levels of PCDD/Fs in Atmosphere of Guangzhou and Typical Emission Sources of PCDD/Fs, China [D]. Guangzhou Institute of Geochemistry, Chinese Academy of Sciences, 2007.

Yuan X, et al. Characteristics and Environmental Effects of Polychlorinated Biphenyls in Sediments From Taihu Lake [J]. *Resources and Environment in the Yangtze Basin,* 2004,13(3): 272-276.

Yun X, et al. Present Situation and Distribution Characteristics of President Organic Pollutants in Freshwater in China [J]. *Environmental Chemistry,* 2013,11: 2072-2081.

Zhang A, etc. Content and Characteristics of PAHs and alkanes in Air Pollution Source of Guangzhou City [J*]. Journal of Anhui Agriculture Science,* 2010, 38(24): 13158-13161.

Zhang J. Distribution Characteristics and Source of Polycyclic Aromatic Hydrocarbons in Various Functional Areas of Shanghai [D]. Shanghai Jiao Tong University, 2008.

Zhang Q, etc. Investigation of Polycyclic Aromatic Hydrocarbons in Atmospheric PM_{10} in Nanjing City[J]. *Environmental Science and Technology,* 2007,10:42-44+117-118.

Zhang Xiaoling, Process and Measure of Dioxin Emission Reduction in Developed Countries [J]. *Environment and Ecology in the Three Gorges,* 2013(1):36.

Zhang Z. Polychlorinated Biphenyls in Chinese Air and Surface Soil: Spatial Distribution Characteristics and Their Inherent Causes[J]. Harbin Institute of Technology, 2010.

Zhao Yanyan, Wang Ling, Lou Yinghua, et al. Determination of a novel bromation flame retardant in the intertidal sediments on the north coast of Jiaozhou Bay by GC-MS[J]. *Environmental Chemistry*, 2015(2).

Zheng, GJ, et al. Polychlorinated dibenzo-p-dioxins and dibenzofurans pollution in China: sources, environmental levels and potential human health impacts.[J]. *Environment International*, 2008, 34(7):1050–1061.

Zhou B, etc. Seasonal Variation and Health Risk Assessment of Atmospheric Polycyclic Aromatic Hydrocarbons (PAHs) in the Urban Area of Xi'an[J]. *Acta Scientiae Circumstantiae*. 2012, 32(9): 2324-2331.

Zhou Bingyan, etc. Oversea Environmentally Sound Management of Persistent Organic Pollutant Wastes [J], *Environmental Protection of Chemical Industry,* 2006(5):430.

Zhou Z, etc. Preliminary Study on Level and Pattern of PCDD/Fs in Soil in Part of Beijing City [J]. *Research of Environmental Sciences,* 2006,(6):54-58.

Zhu Y, Characteristics, Sources Apportionment and Ecological Risk Assessment of Polycyclic Aromatic Hydrocarbons in Soils of Tianjin, China [J]. 2014, 33(2):249-255.

AUTHORS' CONTACT INFORMATION

Dr. Yuanyuan Zeng,
Ph.D. candidate of
Law School of Renmin University of China,
Visiting scholar of Robert H. McKinney School
of Law of Indiana University
Email: zengyuan6652@sina.com

Qian Zeng
Law School, Xiangtan University of China
Email: 1936419051@qq.com

Professor Ke Zhou,
Dean of Research Center of Environmental and Resources Law
Law School of Renmin University of China,
Email: rdzhouke@x263.net

INDEX

A

B

C

D

E

F

G

M

N

O

P

R

S

T

U